Graduate Texts i

Richard H. Crowell
Ralph H. Fox

Introduction to Knot Theory

Springer-Verlag
New York Heidelberg Berlin

R. H. Crowell
Department of Mathematics
Dartmouth College
Hanover, New Hampshire 03755

R. H. Fox
Formerly of Princeton University
Princeton, New Jersey

AMS Subject Classifications: 20E40, 55A05, 55A25, 55A30

Library of Congress Cataloging in Publication Data
Crowell, Richard H.
Introduction to knot theory.
(Graduate texts in mathematics ; 57)
Bibliography: p.
Includes index.
1. Knot theory. I. Fox, Ralph Hartzler, 1913– joint author. II. Title. III. Series.
QA612.2.C76 1977 514′.224 77-22776
ISBN

Printed in the United States of America

9 8 7 6 5 4

ISBN 0-387-90272-4 Springer-Verlag New York

ISBN 3-540-90272-4 Springer-Verlag Berlin Heidelberg

To the memory of

Richard C. Blanchfield and Roger H. Kyle

and

RALPH H. FOX

Preface to the Springer Edition

This book was written as an introductory text for a one-semester course and, as such, it is far from a comprehensive reference work. Its lack of completeness is now more apparent than ever since, like most branches of mathematics, knot theory has expanded enormously during the last fifteen years. The book could certainly be rewritten by including more material and also by introducing topics in a more elegant and up-to-date style. Accomplishing these objectives would be extremely worthwhile. However, a significant revision of the original work along these lines, as opposed to writing a new book, would probably be a mistake. As inspired by its senior author, the late Ralph H. Fox, this book achieves qualities of effectiveness, brevity, elementary character, and unity. These characteristics would be jeopardized, if not lost, in a major revision. As a result, the book is being republished unchanged, except for minor corrections. The most important of these occurs in Chapter III, where the old sections 2 and 3 have been interchanged and somewhat modified. The original proof of the theorem that a group is free if and only if it is isomorphic to $F[\mathscr{A}]$ for some alphabet $\mathscr{A}$ contained an error, which has been corrected using the fact that equivalent reduced words are equal.

I would like to include a tribute to Ralph Fox, who has been called the father of modern knot theory. He was indisputably a first-rate mathematician of international stature. More importantly, he was a great human being. His students and other friends respected him, and they also loved him. This edition of the book is dedicated to his memory.

Richard H. Crowell

Dartmouth College
1977

Preface

Knot theory is a kind of geometry, and one whose appeal is very direct because the objects studied are perceivable and tangible in everyday physical space. It is a meeting ground of such diverse branches of mathematics as group theory, matrix theory, number theory, algebraic geometry, and differential geometry, to name some of the more prominent ones. It had its origins in the mathematical theory of electricity and in primitive atomic physics, and there are hints today of new applications in certain branches of chemistry.[1] The outlines of the modern topological theory were worked out by Dehn, Alexander, Reidemeister, and Seifert almost thirty years ago. As a subfield of topology, knot theory forms the core of a wide range of problems dealing with the position of one manifold imbedded within another.

This book, which is an elaboration of a series of lectures given by Fox at Haverford College while a Philips Visitor there in the spring of 1956, is an attempt to make the subject accessible to everyone. Primarily it is a textbook for a course at the junior-senior level, but we believe that it can be used with profit also by graduate students. Because the algebra required is not the familiar commutative algebra, a disproportionate amount of the book is given over to necessary algebraic preliminaries. However, this is all to the good because the study of noncommutativity is not only essential for the development of knot theory but is itself an important and not overcultivated field. Perhaps the most fascinating aspect of knot theory is the interplay between geometry and this noncommutative algebra.

For the past thirty years Kurt Reidemeister's Ergebnisse publication *Knotentheorie* has been virtually the only book on the subject. During that time many important advances have been made, and moreover the combinatorial point of view that dominates *Knotentheorie* has generally given way to a strictly topological approach. Accordingly, we have emphasized the topological invariance of the theory throughout.

There is no doubt whatever in our minds but that the subject centers around the concepts: *knot group*, *Alexander matrix*, *covering space*, and our presentation is faithful to this point of view. We regret that, in the interest of keeping the material at as elementary a level as possible, we did not introduce and make systematic use of covering space theory. However, had we done so, this book would have become much longer, more difficult, and

[1] H.L. Frisch and E. Wasserman, "Chemical Topology," *J. Am. Chem. Soc.*, 83 (1961) 3789–3795

presumably also more expensive. For the mathematician with some maturity, for example one who has finished studying this book, a survey of this central core of the subject may be found in Fox's "A quick trip through knot theory" (1962).[1]

The bibliography, although not complete, is comprehensive far beyond the needs of an introductory text. This is partly because the field is in dire need of such a bibliography and partly because we expect that our book will be of use to even sophisticated mathematicians well beyond their student days. To make this bibliography as useful as possible, we have included a *guide to the literature*.

Finally, we thank the many mathematicians who had a hand in reading and criticizing the manuscript at the various stages of its development. In particular, we mention Lee Neuwirth, J. van Buskirk, and R. J. Aumann, and two Dartmouth undergraduates, Seth Zimmerman and Peter Rosmarin. We are also grateful to David S. Cochran for his assistance in updating the bibliography for the third printing of this book.

---

[1] See Bibliography

Contents

Chapter VI · Presentation of a Knot Group

Chapter VII · The Free Calculus and the Elementary Ideals

Chapter VIII · The Knot Polynomials

Chapter IX · Characteristic Properties of the Knot Polynomials

Prerequisites

For an intelligent reading of this book a knowledge of the elements of modern algebra and point-set topology is sufficient. Specifically, we shall assume that the reader is familiar with the concept of a function (or mapping) and the attendant notions of domain, range, image, inverse image, one-one, onto, composition, restriction, and inclusion mapping; with the concepts of equivalence relation and equivalence class; with the definition and elementary properties of open set, closed set, neighborhood, closure, interior, induced topology, Cartesian product, continuous mapping, homeomorphism, compactness, connectedness, open cover(ing), and the Euclidean n-dimensional space R^n; and with the definition and basic properties of homomorphism, automorphism, kernel, image, groups, normal subgroups, quotient groups, rings, (two-sided) ideals, permutation groups, determinants, and matrices. These matters are dealt with in many standard textbooks. We may, for example, refer the reader to A. H. Wallace, *An Introduction to Algebraic Topology* (Pergamon Press, 1957), Chapters I, II, and III, and to G. Birkhoff and S. MacLane, *A Survey of Modern Algebra*, Revised Edition (The Macmillan Co., New York, 1953), Chapters III, §§1-3, 7, 8; VI, §§4-8, 11-14; VII, §5; X, §§1, 2; XIII, §§1-4. Some of these concepts are also defined in the index.

In Appendix I an additional requirement is a knowledge of differential and integral calculus.

The usual set theoretic symbols $\in$, $\subset$, $\supset$, $=$, $\cup$, $\cap$, and $-$ are used. For the inclusion symbol we follow the common convention: $A \subset B$ means that $p \in B$ whenever $p \in A$. For the image and inverse image of A under f we write either fA and $f^{-1}A$, or $f(A)$ and $f^{-1}(A)$. For the restriction of f to A we write $f \mid A$, and for the composition of two mappings $f\colon X \to Y$ and $g\colon Y \to Z$ we write gf.

When several mappings connecting several sets are to be considered at the same time, it is convenient to display them in a (mapping) diagram, such as

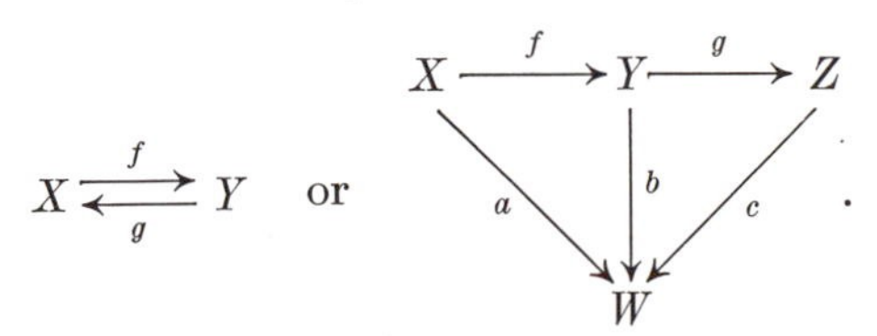

If each element in each set displayed in a diagram has at most one image element in any given set of the diagram, the diagram is said to be *consistent*.

Thus the first diagram is consistent if and only if $gf = 1$ and $fg = 1$, and the second diagram is consistent if and only if $bf = a$ and $cg = b$ (and hence $cgf = a$).

The reader should note the following "diagram-filling" lemma, the proof of which is straightforward.

If h: $G \to H$ and k: $G \to K$ are homomorphisms and h is onto, there exists a (necessarily unique) homomorphism f: $H \to K$ making the diagram

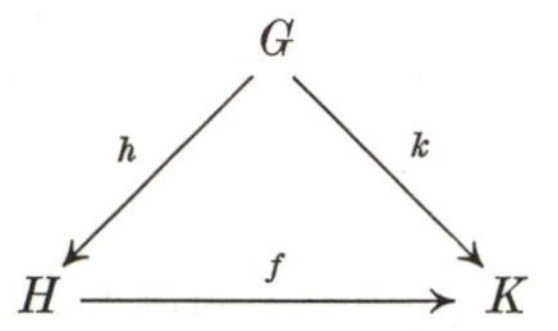

consistent if and only if the kernel of h is contained in the kernel of k.

CHAPTER I

Knots and Knot Types

1. Definition of a knot. Almost everyone is familiar with at least the simplest of the common knots, e.g., the overhand knot, Figure 1, and the figure-eight knot, Figure 2. A little experimenting with a piece of rope will convince anyone that these two knots are different: one cannot be transformed into the other without passing a loop over one of the ends, i.e.,without "tying" or "untying." Nevertheless, failure to change the figure-eight into the overhand by hours of patient twisting is no proof that it can't be done. The problem that we shall consider is the problem of showing mathematically that these knots (and many others) are distinct from one another.

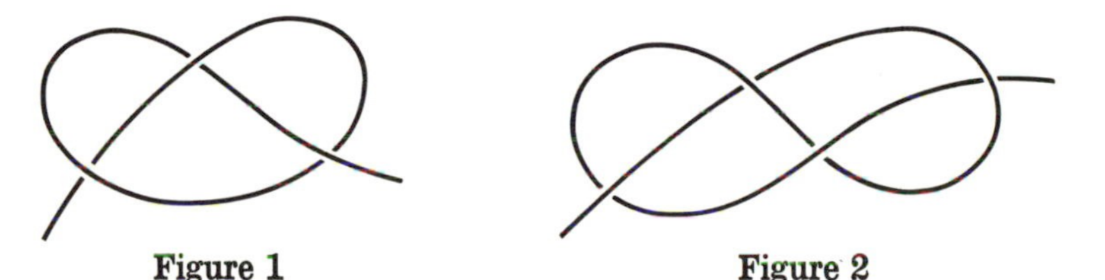

Figure 1 **Figure 2**

Mathematics never proves anything about anything except mathematics, and a piece of rope is a physical object and not a mathematical one. So before worrying about proofs, we must have a mathematical definition of what a knot is and another mathematical definition of when two knots are to be considered the same. This problem of formulating a mathematical model arises whenever one applies mathematics to a physical situation. The definitions should define mathematical objects that approximate the physical objects under consideration as closely as possible. The model may be good or bad according as the correspondence between mathematics and reality is good or bad. There is, however, no way to prove (in the mathematical sense, and it is probably only in this sense that the word has a precise meaning) that the mathematical definitions describe the physical situation exactly.

Obviously, the figure-eight knot can be transformed into the overhand knot by tying and untying—in fact all knots are equivalent if this operation is allowed. Thus tying and untying must be prohibited either in the definition of when two knots are to be considered the same or from the beginning in the very definition of what a knot is. The latter course is easier and is the one we shall adopt. Essentially, we must get rid of the ends. One way would be to prolong the ends to infinity; but a simpler method is to splice them together. Accordingly, we shall consider a knot to be a subset of 3-dimensional space which is homeomorphic to a circle. The formal definition is: K is a *knot* if there exists a homeomorphism of the unit circle C into 3-dimensional space R^3 whose image is K. By the circle C is meant the set of points (x,y) in the plane R^2 which satisfy the equation $x^2 + y^2 = 1$.

The overhand knot and the figure-eight knot are now pictured as in Figure 3 and Figure 4. Actually, in this form the overhand knot is often called the *clover-leaf knot*. Another common name for this knot is the *trefoil*. The figure-eight knot has been called both the *four-knot* and *Listing's knot*.

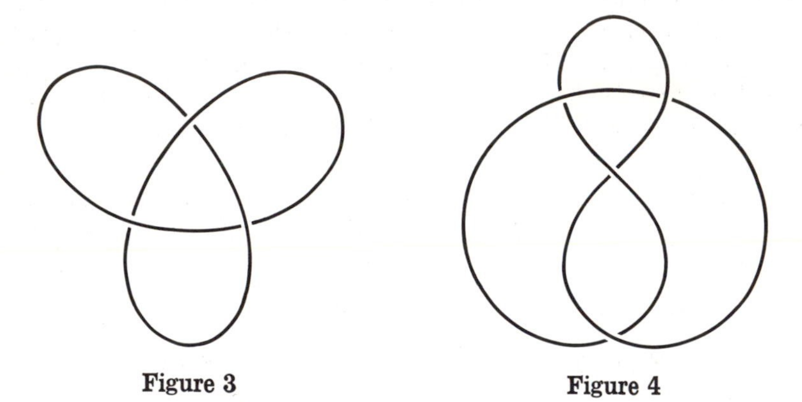

Figure 3 **Figure 4**

We next consider the question of when two knots K_1 and K_2 are to be considered the same. Notice, first of all, that this is not a question of whether or not K_1 and K_2 are homeomorphic. They are both homeomorphic to the unit circle and, consequently, to each other. The property of being knotted is not an intrinsic topological property of the space consisting of the points of the knot, but is rather a characteristic of the way in which that space is imbedded in R^3. Knot theory is a part of 3-dimensional topology and not of 1-dimensional topology. If a piece of rope in one position is twisted into another, the deformation does indeed determine a one-one correspondence between the points of the two positions, and since cutting the rope is not allowed, the correspondence is bicontinuous. In addition, it is natural to think of the motion of the rope as accompanied by a motion of the surrounding air molecules which thus determines a bicontinuous permutation of the points of space. This picture suggests the definition: Knots K_1 and K_2 are *equivalent* if there exists a homeomorphism of R^3 onto itself which maps K_1 onto K_2.

It is a triviality that the relation of knot equivalence is a true equivalence relation. Equivalent knots are said to be of the same *type*, and each equivalence class of knots is a *knot type*. Those knots equivalent to the unknotted circle $x^2 + y^2 = 1$, $z = 0$, are called *trivial* and constitute the *trivial type*.[1] Similarly, the type of the clover-leaf knot, or of the figure-eight knot is defined as the equivalence class of some particular representative knot. The informal statement that the clover-leaf knot and the figure-eight knot are different is rigorously expressed by saying that they belong to distinct knot types.

2. Tame versus wild knots. A *polygonal knot* is one which is the union of a finite number of closed straight-line segments called *edges*, whose endpoints are the *vertices* of the knot. A knot is *tame* if it is equivalent to a polygonal knot; otherwise it is *wild*. This distinction is of fundamental importance. In fact, most of the knot theory developed in this book is applicable (as it stands) only to tame knots. The principal invariants of knot type, namely, the elementary ideals and the knot polynomials, are not necessarily defined for a wild knot. Moreover, their evaluation is based on finding a polygonal representative to start with. The discovery that knot theory is largely confined to the study of polygonal knots may come as a surprise—especially to the reader who approaches the subject fresh from the abstract generality of point-set topology. It is natural to ask what kinds of knots other than polygonal are tame. A partial answer is given by the following theorem.

(2.1) *If a knot parametrized by arc length is of class* C^1 (*i.e., is continuously differentiable*), *then it is tame.*

A proof is given in Appendix I. It is complicated but straightforward, and it uses nothing beyond the standard techniques of advanced calculus. More explicitly, the assumptions on K are that it is rectifiable and given as the image of a vector-valued function $p(s) = (x(s), y(s), z(s))$ of arc length s with continuous first derivatives. Thus, every sufficiently smooth knot is tame.

It is by no means obvious that there exist any wild knots. For example, no knot that lies in a plane is wild. Although the study of wild knots is a corner of knot theory outside the scope of this book, Figure 5 gives an example of a knot known to be wild.[2] This knot is a remarkable curve. Except for the fact that the number of loops increases without limit while their size decreases without limit (as is indicated in the figure by the dotted square about p), the

---

[1] Any knot which lies in a plane is necessarily trivial. This is a well-known and deep theorem of plane topology. See A. H. Newman, *Elements of the Topology of Plane Sets of Points*, Second edition (Cambridge University Press, Cambridge, 1951), p. 173.

[2] R. H. Fox, "A Remarkable Simple Closed Curve," *Annals of Mathematics*, Vol. 50 (1949), pp. 264, 265.

Figure 5

knot could obviously be untied. Notice also that, except at the single point p, it is as smooth and differentiable as we like.

3. Knot projections. A knot K is usually specified by a projection; for example, Figure 3 and Figure 4 show projected images of the clover-leaf knot and the figure-eight knot, respectively. Consider the parallel projection

$$\mathscr{P}: R^3 \to R^3$$

defined by $\mathscr{P}(x,y,z) = (x,y,0)$. A point p of the image $\mathscr{P}K$ is called a *multiple point* if the inverse image $\mathscr{P}^{-1}p$ contains more than one point of K. The *order* of $p \in \mathscr{P}K$ is the cardinality of $(\mathscr{P}^{-1}p) \cap K$. Thus, a *double point* is a multiple point of order 2, a *triple point* is one of order 3, and so on. Multiple points of infinite order can also occur. In general, the image $\mathscr{P}K$ may be quite complicated in the number and kinds of multiple points present. It is possible, however, that K is equivalent to another knot whose projected image is fairly simple. For a polygonal knot, the criterion for being fairly simple is that the knot be in what is called regular position. The definition is as follows: a polygonal knot K is in *regular position* if: (i) the only multiple points of K are double points, and there are only a finite number of them; (ii) no double point is the image of any vertex of K. The second condition insures that every double point depicts a genuine crossing, as in Figure 6a. The sort of double point shown in Figure 6b is prohibited.

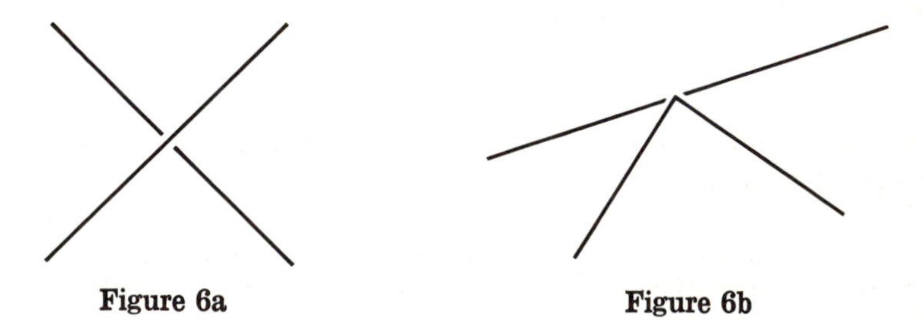

Figure 6a Figure 6b

Each double point of the projected image of a polygonal knot in regular position is the image of two points of the knot. The one with the larger z-coordinate is called an *overcrossing*, and the other is the corresponding *undercrossing*.

(3.1) *Any polygonal knot K is equivalent under an arbitrarily small rotation of R^3 to a polygonal knot in regular position.*

Proof. The geometric idea is to hold K fixed and move the projection. Every bundle (or pencil) of parallel lines in R^3 determines a unique parallel projection of R^3 onto the plane through the origin perpendicular to the bundle. We shall assume the obvious extension of the above definition of regular position so that it makes sense to ask whether or not K is in regular position with respect to any parallel projection. It is convenient to consider R^3 as a subset[3] of a real projective 3-space P^3. Then, to every parallel projection we associate the point of intersection of any line parallel to the direction of projection with the projective plane P^2 at infinity. This correspondence is clearly one-one and onto. Let Q be the set of all points of P^2 corresponding to projections with respect to which K is not in regular position. We shall show that Q is nowhere dense in P^2. It then follows that there is a projection $\mathscr{P}_0$ with respect to which K is in regular position and which is arbitrarily close to the original projection $\mathscr{P}$ along the z-axis. Any rotation of R^3 which transforms the line $\mathscr{P}_0^{-1}(0,0,0)$ into the z-axis will suffice to complete the proof.

In order to prove that Q is nowhere dense in P^2, consider first the set of all straight lines which join a vertex of K to an edge of K. These intersect P^2 in a finite number of straight-line segments whose union we denote by Q_1. Any projection corresponding to a point of $P^2 - Q_1$ must obviously satisfy condition (ii) of the definition of regular position. Furthermore, it can have at most a finite number of multiple points, no one of which is of infinite order. It remains to show that multiple points of order $n \geq 3$ can be avoided, and this is done as follows. Consider any three mutually skew straight lines, each of which contains an edge of K. The locus of all straight lines which intersect these three is a quadric surface which intersects P^2 in a conic section. (See the reference in the preceding footnote.) Set Q_2 equal to the union of all such conics. Obviously, there are only a finite number of them. Furthermore, the image of K under any projection which corresponds to some point of $P^2 - (Q_1 \cup Q_2)$ has no multiple points of order $n \geq 3$. We have shown that

$$P^2 - (Q_1 \cup Q_2) \subset P^2 - Q.$$

Thus Q is a subset of $Q_1 \cup Q_2$, which is nowhere dense in P^2. This completes the proof of (3.1).

[3] For an account of the concepts used in this proof, see O. Veblen and J. W. Young, *Projective Geometry* (Ginn and Company, Boston, Massachusetts, 1910), Vol. 1 pp. 11, 299, 301.

Thus, every tame knot is equivalent to a polygonal knot in regular position. This fact is the starting point for calculating the basic invariants by which different knot types are distinguished.

4. Isotopy type, amphicheiral and invertible knots. This section is not a prerequisite for the subsequent development of knot theory in this book. The contents are nonetheless important and worth reading even on the first time through.

Our definition of knot type was motivated by the example of a rope in motion from one position in space to another and accompanied by a displacement of the surrounding air molecules. The resulting definition of equivalence of knots abstracted from this example represents a simplification of the physical situation, in that no account is taken of the motion during the transition from the initial to the final position. A more elaborate construction, which does model the motion, is described in the definition of the isotopy type of a knot. An *isotopic deformation* of a topological space X is a family of homeomorphisms h_t, $0 \leq t \leq 1$, of X onto itself such that h_0 is the identity, i.e., $h_0(p) = p$ for all p in X, and the function H defined by $H(t,p) = h_t(p)$ is simultaneously continuous in t and p. This is a special case of the general definition of a deformation which will be studied in Chapter V. Knots K_1 and K_2 are said to belong to the same *isotopy type* if there exists an isotopic deformation $\{h_t\}$ of R^3 such that $h_1K_1 = K_2$. The letter t is intentionally chosen to suggest time. Thus, for a fixed point $p \in R^3$, the point $h_t(p)$ traces out, so to speak, the path of the molecule originally at p during the motion of the rope from its initial position at K_1 to K_2.

Obviously, if knots K_1 and K_2 belong to the same isotopy type, they are equivalent. The converse, however, is false. The following discussion of orientation serves to illustrate the difference between the two definitions.

Every homeomorphism h of R^3 onto itself is either *orientation preserving* or *orientation reversing*. Although a rigorous treatment of this concept is usually given by homology theory,[4] the intuitive idea is simple. The homeomorphism h preserves orientation if the image of every right (left)-hand screw is again a right (left)-hand screw; it reverses orientation if the image of every right (left)-hand screw is a left (right)-hand screw. The reason that there is no other possibility is that, owing to the continuity of h, the set of points of R^3 at which the twist of a screw is preserved by h is an open set and the same is true of the set of points at which the twist is reversed. Since h is a homeo-

[4] A homeomorphism k of the n-sphere S^n, $n \geq 1$, onto itself is *orientation preserving* or *reversing* according as the isomorphism $k_*: H_n(S^n) \rightarrow H_n(S^n)$ is or is not the identity. Let $S^n = R^n \cup \{\infty\}$ be the one point compactification of the real Cartesian n-space R^n. Any homeomorphism h of R^n onto itself has a unique extention to a homeomorphism k of $S^n = R^n \cup \{\infty\}$ onto itself defined by $k \mid R^n = h$ and $k(\infty) = \infty$. Then, h is *orientation preserving* or *reversing* according as k is orientation preserving or reversing.

morphism, every point of R^3 belongs to one of these two disjoint sets; and since R^3 is connected, it follows that one of the two sets is empty. The composition of homeomorphisms follows the usual rule of parity:

h_1	h_2	h_1h_2
preserving	preserving	preserving
reversing	preserving	reversing
preserving	reversing	reversing
reversing	reversing	preserving

Obviously, the identity mapping is orientation preserving. On the other hand, the reflection $(x,y,z) \rightarrow (x,y,-z)$ is orientation reversing. If h is a linear transformation, it is orientation preserving or reversing according as its determinant is positive or negative. Similarly, if both h and its inverse are C^1 differentiable at every point of R^3, then h preserves or reverses orientation according as its Jacobian is everywhere positive or everywhere negative.

Consider an isotopic deformation $\{h_t\}$ of R^3. The fact that the identity is orientation preserving combined with the continuity of $H(t,p) = h_t(p)$, suggests that h_t is orientation preserving for every t in the interval $0 \leq t \leq 1$. This is true.[5] As a result, we have that a necessary condition for two knots to be of the same isotopy type is that there exist an orientation preserving homeomorphism of R^3 on itself which maps one knot onto the other.

A knot K is said to be *amphicheiral* if there exists an orientation reversing homeomorphism h of R^3 onto itself such that $hK = K$. An equivalent formulation of the definition, which is more appealing geometrically, is provided by the following lemma. By the *mirror image* of a knot K we shall mean the image of K under the reflection $\mathscr{R}$ defined by $(x,y,z) \rightarrow (x,y,-z)$. Then,

(4.1) *A knot K is amphicheiral if and only if there exists an orientation preserving homeomorphism of R^3 onto itself which maps K onto its mirror image.*

Proof. If K is amphicheiral, the composition $\mathscr{R}h$ is orientation preserving and maps K onto its mirror image. Conversely, if h' is an orientation preserving homeomorphism of R^3 onto itself which maps K onto its mirror image, the composition $\mathscr{R}h'$ is orientation reversing and $(\mathscr{R}h')K = K$.

It is not hard to show that the figure-eight knot is amphicheiral. The experimental approach is the best; a rope which has been tied as a figure-eight and then spliced is quite easily twisted into its mirror image. The operation is illustrated in Figure 7. On the other hand, the clover-leaf knot is not amphi-

[5] Any isotopic deformation $\{h_t\}$, $0 \leq t \leq 1$, of the Cartesian n-space R^n definitely possesses a unique extension to an isotopic deformation $\{k_t\}$, $0 \leq t \leq 1$, of the n-sphere S^n, i.e., $k_t \mid R^n = h_t$, and $k_t(\infty) = \infty$. For each t, the homeomorphism k_t is homotopic to the identity, and so the induced isomorphism $(k_t)_*$ on $H_n(S^n)$ is the identity. It follows that h_t is orientation preserving for all t in $0 \leq t \leq 1$. (See also footnote 4.)

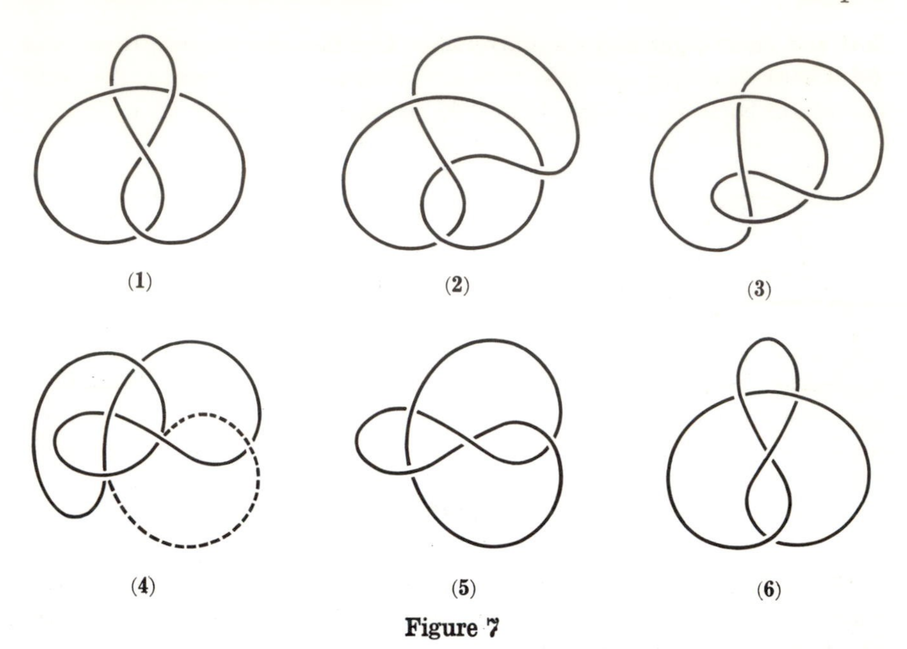

Figure 7

cheiral. In this case, experimenting with a piece of rope accomplishes nothing except possibly to convince the skeptic that the question is nontrivial. Actually, to prove that the clover-leaf is not amphicheiral is hard and requires fairly advanced techniques of knot theory. Assuming this result, however, we have that the clover-leaf knot and its mirror image are equivalent but not of the same isotopy type.

It is natural to ask whether or not every orientation preserving homeomorphism f of R^3 onto itself is realizable by an isotopic deformation, i.e., given f, does there exist $\{h_t\}$, $0 \leq t \leq 1$, such that $f = h_1$? If the answer were no, we would have a third kind of knot type. This question is not an easy one. The answer is, however, yes.[6]

Just as every homeomorphism of R^3 onto itself either preserves or reverses orientation, so does every homeomorphism f of a knot K onto itself. The geometric interpretation is analogous to, and simpler than, the situation in 3-dimensional space. Having prescribed a direction on the knot, f preserves or reverses orientation according as the order of points of K is preserved or reversed under f. A knot K is called *invertible* if there exists an orientation preserving homeomorphism h of R^3 onto itself such that the restriction $h \mid K$ is an orientation reversing homeomorphism of K onto itself. Both the clover-

[6] G. M. Fisher, "On the Group of all Homeomorphisms of a Manifold," *Transactions of the American Mathematical Society,* Vol. 97 (1960), pp. 193–212.

leaf and figure-eight knots are invertible. One has only to turn them over (cf. Figure 8).

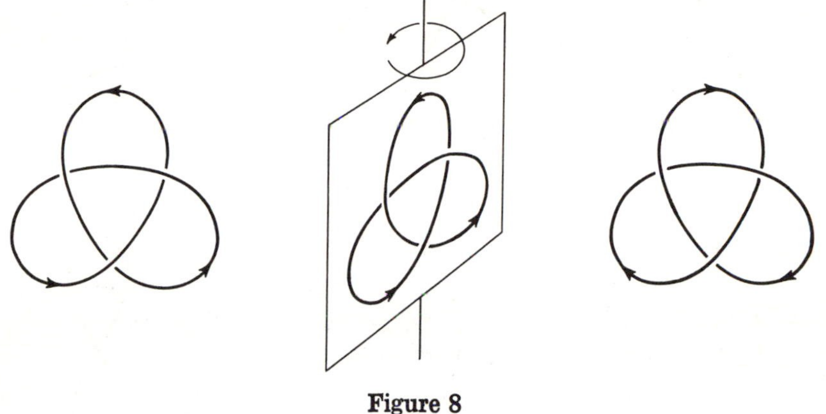

Figure 8

Until recently no example of a noninvertible knot was known. Trotter solved the problem by exhibiting an infinite set of noninvertible knots, one of which is shown in Figure 9.[7]

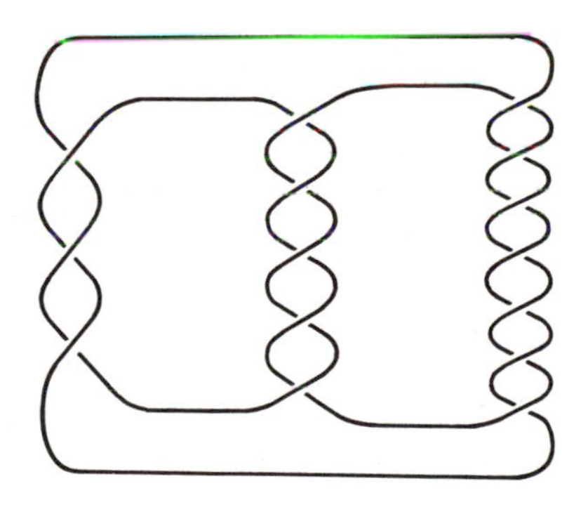

Figure 9

EXERCISES

1. Show that any simple closed polygon in R^2 belongs to the trivial knot type.

2. Show that there are no knotted quadrilaterals or pentagons. What knot types are represented by hexagons? by septagons?

---

[7] H. F. Trotter, "Noninvertible knots exist." *Topology*, vol. 2 (1964), pp. 275–280.

3. Devise a method for constructing a table of knots, and use it to find the ten knots of not more than six crossings. (Do not consider the question of whether these are really distinct types.)

4. Determine by experiment which of the above ten knots are obviously amphicheiral, and verify that they are all invertible.

5. Show that the number of tame knot types is at most countable.

6. (Brunn) Show that any knot is equivalent to one whose projection has at most one multiple point (perhaps of very high order).

7. (Tait) A polygonal knot in regular position is said to be *alternating* if the undercrossings and overcrossings alternate around the knot. (A knot type is called alternating if it has an alternating representative.) Show that if K is any knot in regular position there is an alternating knot (in regular position) that has the same projection as K.

8. Show that the regions into which R^2 is divided by a regular projection can be colored black and white in such a way that adjacent regions are of opposite colors (as on a chessboard).

9. Prove the assertion made in footnote 4 that any homeomorphism h of R^n onto itself has a unique extension to a homeomorphism k of $S^n = R^n \cup \{\infty\}$ onto itself.

10. Prove the assertion made in footnote 5 that any isotopic deformation $\{h_t\}$, $0 \leq t \leq 1$, of R^n possesses a unique extension to an isotopic deformation $\{k_t\}$, $0 \leq t \leq 1$, of S^n. (*Hint:* Define $F(p, t) = (h_t(p), t)$, and use *invariance of domain* to prove that F is a homeomorphism of $R^n \times [0, 1]$ onto itself.)

CHAPTER II

The Fundamental Group

Introduction. Elementary analytic geometry provides a good example of the applications of formal algebraic techniques to the study of geometric concepts. A similar situation exists in algebraic topology, where one associates algebraic structures with the purely topological, or geometric, configurations. The two basic geometric entities of topology are topological spaces and continuous functions mapping one space into another. The algebra involved, in contrast to that of ordinary analytic geometry, is what is frequently called modern algebra. To the spaces and continuous maps between them are made to correspond groups and group homomorphisms. The analogy with analytic geometry, however, breaks down in one essential feature. Whereas the coordinate algebra of analytic geometry completely reflects the geometry, the algebra of topology is only a partial characterization of the topology. This means that a typical theorem of algebraic topology will read: If topological spaces X and Y are homeomorphic, then such and such algebraic conditions are satisfied. The converse proposition, however, will generally be false. Thus, if the algebraic conditions are not satisfied, we know that X and Y are topologically distinct. If, on the other hand, they are fulfilled, we usually can conclude nothing. The bridge from topology to algebra is almost always a one-way road; but even with that one can do a lot.

One of the most important entities of algebraic topology is the fundamental group of a topological space, and this chapter is devoted to its definition and elementary properties. In the first chapter we discussed the basic spaces and continuous maps of knot theory: the 3-dimensional space R^3, the knots themselves, and the homeomorphisms of R^3 onto itself which carry one knot onto another of the same type. Another space of prime importance is the *complementary space* $R^3 - K$ of a knot K, which consists of all of those points of R^3 that do not belong to K. All of the knot theory in this book is a study of the properties of the fundamental groups of the complementary spaces of knots, and this is indeed the central theme of the entire subject. In this chapter, however, the development of the fundamental group is made for an arbitrary topological space X and is independent of our later applications of the fundamental group to knot theory.

1. Paths and loops. A particle moving in space during a certain interval of time describes a path. It will be convenient for us to assume that the motion begins at time $t = 0$ and continues until some stopping time, which may differ for different paths but may be either positive or zero. For any two real numbers x and y with $x \leq y$, we define $[x,y]$ to be the set of all real numbers t satisfying $x \leq t \leq y$. A *path* a in a topological space X is then a continuous mapping

$$a: [0, \| a \|] \rightarrow X.$$

The number $\| a \|$ is the *stopping time*, and it is assumed that $\| a \| \geq 0$. The points $a(0)$ and $a(\| a \|)$ in X are the *initial point* and *terminal point*, respectively, of the path a.

It is essential to distinguish a path a from the set of image points $a(t)$ in X visited during the interval $[0, \| a \|]$. Different paths may very well have the same set of image points. For example, let X be the unit circle in the plane, given in polar coordinates as the set of all pairs (r,θ) such that $r = 1$. The two paths

$$a(t) = (1,t), \quad 0 \leq t \leq 2\pi,$$
$$b(t) = (1,2t), \quad 0 \leq t \leq 2\pi,$$

are distinct even though they have the same stopping time, same initial and terminal point, and same set of image points. Paths a and b are *equal* if and only if they have the same domain of definition, i.e., $\| a \| = \| b \|$, and, if for every t in that domain, $a(t) = b(t)$.

Consider any two paths a and b in X which are such that the terminal point of a coincides with the initial point of b, i.e., $a(\| a \|) = b(0)$. The *product* $a \cdot b$ of the paths a and b is defined by the formula

$$(a \cdot b)(t) = \begin{cases} a(t), & 0 \leq t \leq \| a \|, \\ b(t - \| a \|), & \| a \| \leq t \leq \| a \| + \| b \|. \end{cases}$$

It is obvious that this defines a continuous function, and $a \cdot b$ is therefore a path in X. Its stopping time is

$$\| a \cdot b \| = \| a \| + \| b \|.$$

We emphasize that the product of two paths is not defined unless the terminal point of the first is the same as the initial point of the second. It is obvious that the three assertions

(i) $a \cdot b$ *and* $b \cdot c$ *are defined,*
(ii) $a \cdot (b \cdot c)$ *is defined,*
(iii) $(a \cdot b) \cdot c$ *is defined,*

are equivalent and that whenever one of them holds, the *associative law,*

$$a \cdot (b \cdot c) = (a \cdot b) \cdot c,$$

is valid.

A path a is called an *identity path*, or simply an identity, if it has stopping time $\| a \| = 0$. This terminology reflects the fact that the set of all identity paths in a topological space may be characterized as the set of all multiplicative identities with respect to the product. That is, *the path e is an identity if and only if $e \cdot a = a$ and $b \cdot e = b$ whenever $e \cdot a$ and $b \cdot e$ are defined.* Obviously, an identity path has only one image point, and conversely, there is precisely one identity path for each point in the space. We call a path whose image is a single point a *constant path.* Every identity path is constant; but the converse is clearly false.

For any path a, we denote by a^{-1} the *inverse path* formed by traversing a in the opposite direction. Thus,

$$a^{-1}(t) = a(\| a \| - t), \qquad 0 \leq t \leq \| a \|.$$

The reason for adopting this name and notation for a^{-1} will become apparent as we proceed. At present, calling a^{-1} an inverse is a misnomer. It is easy to see that $a \cdot a^{-1}$ is an identity e if and only if $a = e$.

The meager algebraic structure of the set of all paths of a topological space with respect to the product is certainly far from being that of a group. One way to improve the situation algebraically is to select an arbitrary point p in X and restrict our attention to paths which begin and end at p. A path whose initial and terminal points coincide is called a *loop,* its common endpoint is its *basepoint,* and a loop with basepoint p will frequently be referred to as *p-based.* The product of any two p-based loops is certainly defined and is again a p-based loop. Moreover, the identity path at p is a multiplicative identity. These remarks are summarized in the statement that *the set of all p-based loops in X is a semi-group with identity.*

The semi-group of loops is a step in the right direction; but it is not a group. Hence, we consider another approach. Returning to the set of all paths, we shall define in the next section a notion of equivalent paths. We shall then consider a new set, whose elements are the equivalence classes of paths. The fundamental group is obtained as a combination of this construction with the idea of a loop.

2. Classes of paths and loops. A collection of paths h_s in X, $0 \leq s \leq 1$, will be called a *continuous family* of paths if

(i) The stopping time $\| h_s \|$ depends continuously on s.

(ii) The function h defined by the formula $h(s,t) = h_s(t)$ maps the closed region $0 \leq s \leq 1$, $0 \leq t \leq \| h_s \|$ continuously into X.

It should be noted that a function of two variables which is continuous at every point of its domain of definition with respect to each variable is *not* necessarily continuous in both simultaneously. The function f defined on the unit square $0 \leq s \leq 1$, $0 \leq t \leq 1$ by the formula

$$f(s,t) = \begin{cases} 1, & \text{if } s = t = 0, \\ \dfrac{s+t}{\sqrt{s^2 + t^2}}, & \text{otherwise,} \end{cases}$$

is an example. The collection of paths $\{f_s\}$ defined by $f_s(t) = f(s,t)$ is not, therefore, a continuous family.

A *fixed-endpoint family* of paths is a continuous family $\{h_s\}$, $0 \le s \le 1$, such that $h_s(0)$ and $h_s(\| h_s \|)$ are independent of s, i.e., there exist points p and q in X such that $h_s(0) = p$ and $h_s(\| h_s \|) = q$ for all s in the interval $0 \le s \le 1$. The difference between a continuous family and a fixed-endpoint family is illustrated below in Figure 10.

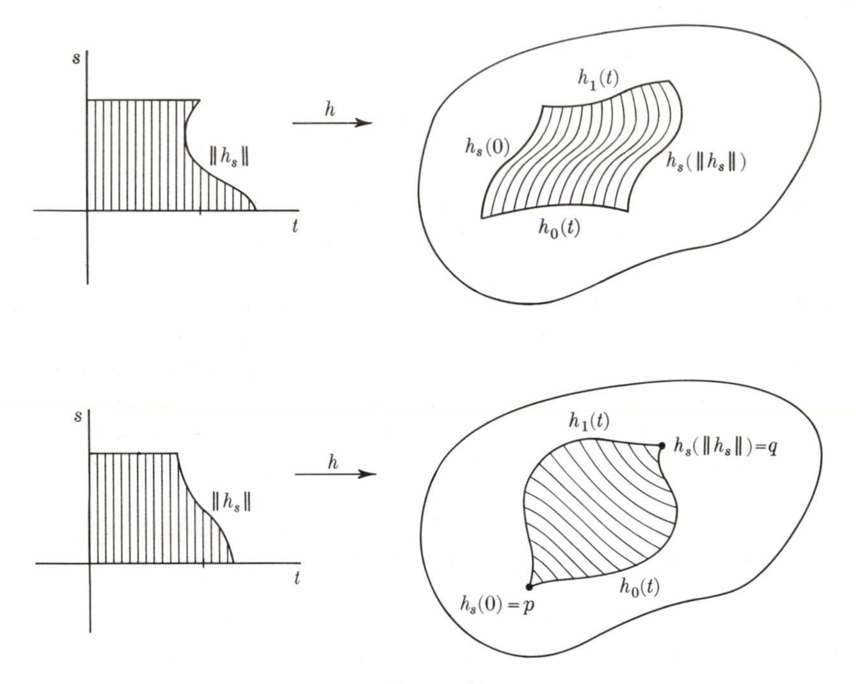

Figure 10

Let a and b be two paths in the topological space X. Then, a is said to be *equivalent* to b, written $a \simeq b$, if there exists a fixed-endpoint family $\{h_s\}$, $0 \le s \le 1$, of paths in X such that $a = h_0$ and $b = h_1$.

The relation $\simeq$ is *reflexive*, i.e., for any path a, we have $a \simeq a$, since we may obviously define $h_s(t) = a(t)$, $0 \le s \le 1$. It is also *symmetric*, i.e., $a \simeq b$ implies $b \simeq a$, because we may define $k_s(t) = h_{1-s}(t)$. Finally, $\simeq$ is *transitive*, i.e., $a \simeq b$ and $b \simeq c$ imply $a \simeq c$. To verify the last statement, let us suppose that h_s and k_s are the fixed-endpoint families exhibiting the equivalences $a \simeq b$ and $b \simeq c$ respectively. Then the collection of paths $\{j_s\}$ defined by

$$j_s(t) = \begin{cases} h_{2s}(t), & 0 \le s \le \frac{1}{2}, \\ k_{2s-1}(t), & \frac{1}{2} \le s \le 1, \end{cases}$$

is a fixed-endpoint family proving $a \simeq c$. To complete the arguments, the reader should convince himself that the collections defined above in showing reflexivity, symmetry, and transitivity actually do satisfy all the conditions for being path equivalences: fixed-endpoint, continuity of stopping time, and simultaneous continuity in s and t.

Thus, the relation $\simeq$ is a true equivalence relation, and the set of all paths in the space X is therefore partitioned into equivalence classes. We denote the equivalence class containing an arbitrary path a by $[a]$. That is, $[a]$ is the set of all paths b in X such that $a \simeq b$. Hence, we have

$$[a] = [b] \quad \textit{if and only if} \quad a \simeq b.$$

The collection of all equivalence classes of paths in the topological space X will be denoted by $\Gamma(X)$. It is called the *fundamental groupoid of* X. The definition of a groupoid as an abstract entity is given in Appendix II.

Geometrically, paths a and b are equivalent if and only if one can be continuously deformed onto the other in X without moving the endpoints. The definition is the formal statement of this intuitive idea. As an example, let X be the annular region of the plane shown in Figure 11 and consider five loops e (identity), a_1, a_2, a_3, a_4 in X based at p. We have the following equivalences

$$a_1 \simeq a_2 \simeq e,$$
$$a_3 \simeq a_4.$$

However, it is not true that

$$a_1 \simeq a_3.$$

Figure 11 shows that certain fundamental properties of X are reflected in the equivalence structure of the loops of X. If, for example, the points lying inside the inner boundary of X had been included as a part of X, i.e., if the

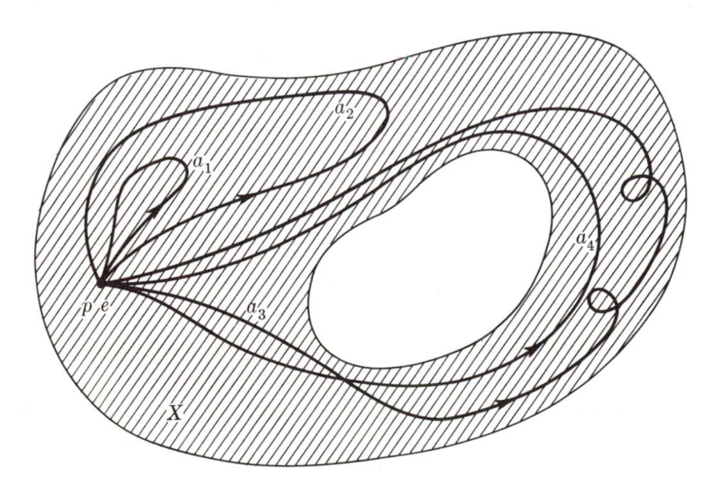

Figure 11

hole were filled in, then all loops based at p would have been equivalent to the identity loop e. It is intended that the arrows in Figure 11 should imply that as the interval of the variable t is traversed for each a_i, the image point runs around the circuit once in the direction of the arrow. It is essential that the idea of a_i as a function be maintained. The image points of a path do not specify the path completely; for example, $a_3 \neq a_3 \cdot a_3$, and furthermore, we do not even have, $a_3 \simeq a_3 \cdot a_3$.

We shall now show that path multiplication induces a multiplication in the fundamental groupoid $\Gamma(X)$. As a result we shall transfer our attention from paths and products of paths to consideration of equivalence classes of paths and the induced multiplication between these classes. In so doing, we shall obtain the necessary algebraic structure for defining the fundamental group.

(2.1) *For any paths a, a', b, b' in X, if $a \cdot b$ is defined and $a \simeq a'$ and $b \simeq b'$, then $a' \cdot b'$ is defined and $a \cdot b \simeq a' \cdot b'$.*

Proof. If $\{h_s\}$ and $\{k_s\}$ are the fixed-endpoint families exhibiting the equivalences $a \simeq a'$ and $b \simeq b'$, respectively, then the collection of paths $\{h_s \cdot k_s\}$ is a fixed-endpoint family which gives $a \cdot b \simeq a' \cdot b'$. We observe, first of all, that the products $h_s \cdot k_s$, are defined for every s in $0 \leq s \leq 1$ because

$$h_s(\| h_s \|) = h_0(\| h_0 \|) = a(\| a \|) = b(0) = k_0(0) = k_s(0).$$

In particular, $a' \cdot b' = h_1 \cdot k_1$ is defined. It is a straightforward matter to verify that the function $h \cdot k$ defined by

$$(h \cdot k)(s,t) = (h_s \cdot k_s)(t), \qquad 0 \leq s \leq 1, 0 \leq t \leq \| h_s \| + \| k_s \|,$$

is simultaneously continuous in s and t. Since $\| h_s \cdot k_s \| = \| h_s \| + \| k_s \|$ is a continuous function of s, the paths $h_s \cdot k_s$ form a continuous family. We have

$$(h_s \cdot k_s)(0) = h_s(0) = a(0),$$

and

$$(h_s \cdot k_s)(\| h_s \cdot k_s \|) = k_s(\| k_s \|) = b(\| b \|),$$

so that $\{h_s \cdot k_s\}$, $0 \leq s \leq 1$, is a fixed-endpoint family. Since $h_0 \cdot k_0 = a \cdot b$ and $h_1 \cdot k_1 = a' \cdot b'$, the proof is complete.

Consider any two paths a and b in X such that $a \cdot b$ is defined. The product of the equivalence classes $[a]$ and $[b]$ is defined by the formula

$$[a] \cdot [b] = [a \cdot b].$$

Multiplication in $\Gamma(X)$ is well-defined as a result of (2.1).

Since all paths belonging to a single equivalence class have the same initial point and the same terminal point, we may define the initial point and terminal point of an element α in $\Gamma(X)$ to be those of an arbitrary representative path in α. The product $\alpha \cdot \beta$ of two elements α and β in $\Gamma(X)$ is then defined if the terminal point of α coincides with the initial point of β. Since the mapping $a \rightarrow [a]$ is product preserving, the associative law holds in $\Gamma(X)$ whenever the relevant products are defined, exactly as it does for paths.

An element ϵ in $\Gamma(X)$ is an *identity* if it contains an identity path. Just as before, we have that *an element ϵ is an identity if and only if $\epsilon \cdot \alpha = \alpha$ and $\beta \cdot \epsilon = \beta$ whenever $\epsilon \cdot \alpha$ and $\beta \cdot \epsilon$ are defined.* This assertion follows almost trivially from the analogous statement for paths. For, let ϵ be an identity, and suppose that $\epsilon \cdot \alpha$ is defined. Let e be an identity path in ϵ and a a representative path in α. Then, $e \cdot a = a$, and so $\epsilon \cdot \alpha = \alpha$. Similarly, $\beta \cdot \epsilon = \beta$. Conversely, suppose that ϵ is not an identity. To prove that there exists an α such that $\epsilon \cdot \alpha$ is defined and $\epsilon \cdot \alpha \neq \alpha$, select for α the class containing the identity path corresponding to the terminal point of ϵ. Then, $\epsilon \cdot \alpha$ is defined, and, since α is an identity, $\epsilon \cdot \alpha = \epsilon$. Hence, if $\epsilon \cdot \alpha = \alpha$, the class ϵ is an identity, which is contrary to assumption. This completes the proof. We conclude that $\Gamma(X)$ has at least as much algebraic structure as the set of paths in X. The significant thing, of course, is that it has more.

(2.2) *For any path a in X, there exist identity paths e_1 and e_2 such that $a \cdot a^{-1} \simeq e_1$ and $a^{-1} \cdot a \simeq e_2$.*

Proof. The paths e_1 and e_2 are obviously the identities corresponding to the initial and terminal points, respectively, of a. Consider the collection of paths $\{h_s\}$, $0 \leq s \leq 1$, defined by the formula

$$h_s(t) = \begin{cases} a(t), & 0 \leq t \leq s \parallel a \parallel, \\ a(2s \parallel a \parallel - t), & s \parallel a \parallel \leq t \leq 2s \parallel a \parallel. \end{cases}$$

The domain of the mapping h defined by $h(s,t) = h_s(t)$ is the shaded area shown in Figure 12. On the line $t = 0$, i.e., on the s-axis, h is constantly equal to $a(0)$. The same is true along the line $t = 2s \parallel a \parallel$. Hence the paths h_s form a

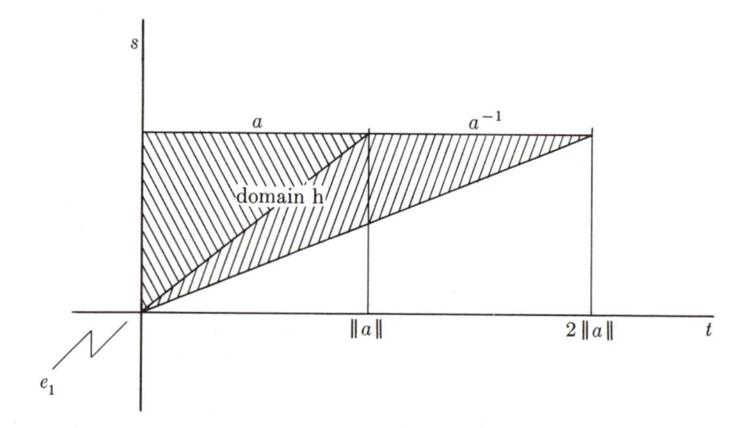

Figure 12

fixed-endpoint family. For values of t along the horizontal line $s = 1$, the function h behaves like $a \cdot a^{-1}$. We have

$$h_0 = e_1$$

$$h_1(t) = \begin{cases} a(t), & 0 \leq t \leq \| a \|, \\ a(2 \| a \| - t), & \| a \| \leq t \leq 2 \| a \|, \end{cases}$$

$$= \begin{cases} a(t), & 0 \leq t \leq \| a \|, \\ a^{-1}(t - \| a \|), & \| a \| \leq t \leq 2 \| a \|, \end{cases}$$

$$= (a \cdot a^{-1})(t),$$

and the proof that $a \cdot a^{-1} \simeq e_1$ is complete. The other equivalence may, of course, be proved in the same way, but it is quicker to use the result just proved to conclude that $a^{-1} \cdot (a^{-1})^{-1} \simeq e_2$. Since $(a^{-1})^{-1} = a$, the proof is complete.

(2.3) *For any paths a and b, if $a \simeq b$, then $a^{-1} \simeq b^{-1}$.*

Proof. This result is a corollary of (2.1) and (2.2). We have

$$a^{-1} \simeq a^{-1} \cdot b \cdot b^{-1} \simeq a^{-1} \cdot a \cdot b^{-1} \simeq b^{-1}.$$

On the basis of (2.3), we define the inverse of an arbitrary element α in $\Gamma(X)$ by the formula

$$\alpha^{-1} = [a^{-1}], \quad \text{for any } a \text{ in } \alpha.$$

The element α^{-1} depends only on α and not on the particular representative path a. That is, α^{-1} is well-defined. This time there is no misnaming. As a corollary of (2.2), we have

(2.4) *For any α in $\Gamma(X)$, there exist identities ϵ_1 and ϵ_2 such that $\alpha \cdot \alpha^{-1} = \epsilon_1$ and $\alpha^{-1} \cdot \alpha = \epsilon_2$.*

The additional abstract property possessed by the fundamental groupoid $\Gamma(X)$ beyond those of the set of all paths in X is expressed in (2.4). We now obtain the fundamental group of X relative to the basepoint p by defining the exact analogue in $\Gamma(X)$ of the p-based loops in the set of all paths: Set $\pi(X,p)$ equal to the subset of $\Gamma(X)$ of all elements having p as both initial and terminal point. The assignment $a \rightarrow [a]$ determines a mapping of the semi-group of p-based loops into $\pi(X,p)$ which is both product preserving and onto. It follows that $\pi(X,p)$ is a semi-group with identity and, by virtue of (2.4), we have

(2.5) *The set $\pi(X,p)$, together with the multiplication defined, is a group.* It is by definition the *fundamental group*[1] *of X relative to the basepoint p.*

[1] The customary notation in topology for this group is $\pi_1(X,p)$. There is a sequence of groups $\pi_n(X,p)$, $n \geq 1$, called the homotopy groups of X relative to p. The fundamental group is the first one of the sequence.

We conclude this section with the useful observation that as far as equivalence classes go, constant paths are the same as identity paths.

(2.6) *Every constant path is equivalent to an identity path.*

Proof. Let k be an arbitrary constant path in X defined by

$$k(t) = p, \quad 0 \leq t \leq \| k \|, \text{ for some } p \in X.$$

Obviously, the collection of paths h_s defined by the formula

$$h_s(t) = p, \quad 0 \leq t \leq s \| k \|$$

is a fixed-endpoint family, and $h_1 = k$ and $h_0 = e$, where e is the identity path corresponding to p.

3. Change of basepoint. The fundamental group $\pi(X,p)$ of X is defined with respect to and depends on the choice of basepoint p. However we shall now show that if X is pathwise connected the fundamental groups of X defined for different basepoints are all isomorphic. A topological space X is *pathwise connected*[2] if any two of its points can be joined by a path lying in X.

(3.1) *Let α be any element of $\Gamma(X)$ having initial point p and terminal point p'. Then, the assignment*

$$\beta \rightarrow \alpha^{-1} \cdot \beta \cdot \alpha \text{ for any } \beta \text{ in } \pi(X,p)$$

is an isomorphism of $\pi(X,p)$ onto $\pi(X,p')$.

Proof. The product $\alpha^{-1} \cdot \beta \cdot \alpha$ is certainly defined, and it is clear that $\alpha^{-1} \cdot \beta \cdot \alpha \in \pi(X,p')$. For any $\beta_1, \beta_2 \in \pi(X,p)$

$$\beta_1 \cdot \beta_2 \rightarrow \alpha^{-1} \cdot (\beta_1 \cdot \beta_2) \cdot \alpha = (\alpha^{-1} \cdot \beta_1 \cdot \alpha) \cdot (\alpha^{-1} \cdot \beta_2 \cdot \alpha).$$

So the mapping is a homomorphism. Next, suppose $\alpha^{-1} \cdot \beta \cdot \alpha = 1$ $(= \epsilon)$. Then,

$$\beta = \alpha \cdot \alpha^{-1} \cdot \beta \cdot \alpha \cdot \alpha^{-1} = \alpha \cdot \alpha^{-1} = 1,$$

and we may conclude that the assignment is an isomorphism. Finally, for any γ in $\pi(X,p')$, $\alpha \cdot \gamma \cdot \alpha^{-1} \in \pi(X,p)$. Obviously,

$$\alpha \cdot \gamma \cdot \alpha^{-1} \rightarrow \gamma.$$

Thus the mapping is onto, and the proof is complete.

---

[2] This definition should be contrasted with that of connectedness.
A topological space is *connected* if it is not the union of two disjoint nonempty open sets. It is easy to show that a pathwise connected space is necessarily connected, but that a connected space is not necessarily pathwise connected.

It is a corollary of (3.1) that the fundamental group of a pathwise connected space is independent of the basepoint in the sense that the groups defined for any two basepoints are isomorphic. For this reason, the definition of the fundamental group is frequently restricted to pathwise connected spaces for which it is customary to omit explicit reference to the basepoint and to speak simply of the fundamental group $\pi(X)$ of X. Occasionally this omission can cause real confusion (if one is interested in properties of $\pi(X,p)$ beyond those it possesses as an abstract group). In any event, $\pi(X)$ always means $\pi(X,p)$ for some choice of basepoint p in X.

4. Induced homomorphisms of fundamental groups. Suppose we are given a continuous mapping $f\colon X \to Y$ from one topological space X into another Y. Any path a in X determines a path fa in Y given by the composition

$$[0, \| a \|] \xrightarrow{a} X \xrightarrow{f} Y,$$

i.e., $fa(t) = f(a(t))$. The stopping time of fa is obviously the same as that of a, i.e., $\| fa \| = \| a \|$. Furthermore, the assignment $a \to fa$ is product-preserving:

(4.1) *If the product $a \cdot b$ is defined, so is $fa \cdot fb$, and $f(a \cdot b) = fa \cdot fb$.*

The proof is very simple. Since $a \cdot b$ is defined, $a(\| a \|) = b(0)$. Consequently,

$$\begin{aligned} fa(\| fa \|) &= fa(\| a \|) = f(a(\| a \|)) \\ &= f(b(0)) = fb(0), \end{aligned}$$

and the product $fa \cdot fb$ is therefore defined. Furthermore,

$$\begin{aligned} f(a \cdot b)(t) &= f((a \cdot b)(t)) \\ &= \begin{cases} f(a(t)), & 0 \le t \le \| a \|, \\ f(b(t - \| a \|)), & \| a \| \le t \le \| a \| + \| b \|, \end{cases} \\ &= \begin{cases} fa(t), & 0 \le t \le \| fa \|, \\ fb(t - \| fa \|), & \| fa \| \le t \le \| fa \| + \| fb \|, \end{cases} \\ &= (fa \cdot fb)(t). \end{aligned}$$

It is obvious that,

(4.2) *If e is an identity, so is fe.*

Furthermore,

(4.3) $$fa^{-1} = (fa)^{-1}.$$

Proof.

$$\begin{aligned} fa^{-1}(t) &= f(a^{-1}(t)) = f(a(\| a \| - t)) \\ &= fa(\| fa \| - t) = (fa)^{-1}(t). \end{aligned}$$

For any continuous family of paths $\{h_s\}$, $0 \leq s \leq 1$, in X, the collection of paths $\{fh_s\}$ is also a continuous family. In addition, $\{fh_s\}$ is a fixed-endpoint family provided $\{h_s\}$ is. Consequently,

(4.4) *If $a \simeq b$, then $fa \simeq fb$.*

Thus, f determines a mapping f_* of the fundamental groupoid $\Gamma(X)$ into the fundamental groupoid $\Gamma(Y)$ given by the formula

$$f_*([a]) = [fa].$$

The basic properties of the function f_* are summarized in

(4.5)

(i) *If ϵ is an identity, then so is $f_*\epsilon$.*

(ii) *If the product $\alpha \cdot \beta$ is defined, then so is $f_*\alpha \cdot f_*\beta$ and $f_*(\alpha \cdot \beta) = f_*\alpha \cdot f_*\beta$.*

(iii) *If f: $X \to X$ is the identity function,* i.e., $f(x) = x$, *then f_* is also the identity function,* i.e., $f_*\alpha = \alpha$.

(iv) *If $X \xrightarrow{f} Y \xrightarrow{g} Z$ are continuous mappings and gf: $X \to Z$ is the composition, then $(gf)_* = g_*f_*$.*

The proofs of these propositions follow immediately from (4.1), (4.2), and the associativity of the composition of functions, i.e., $(gf)a = g(fa)$.

It is obvious that, for any choice of basepoint p in X, $f_*(\pi(X,p)) \subset \pi(Y,fp)$. Thus, the function defined by restricting f_* to $\pi(X,p)$ (which we shall also denote by f_*) determines a homomorphism

$$f_*\colon \pi(X,p) \to \pi(Y,fp),$$

which is called the *homomorphism induced by f*. Notice that if X is pathwise connected, the algebraic properties of the homomorphism f_* are independent of the choice of basepoint. More explicitly, for any two points $p, q \in X$, choose $\alpha \in \Gamma(X)$ with initial point p and terminal point q. Then the homomorphisms

$$\text{(4.6)} \qquad \begin{array}{ccc} \pi(X,p) & \xrightarrow{f_*} & \pi(Y,fp) \\ \beta \longrightarrow \alpha^{-1}\beta\alpha \ \Big\downarrow & & \Big\downarrow \ \gamma \longrightarrow (f_*\alpha)^{-1}\gamma(f_*\alpha) \\ \pi(X,q) & \xrightarrow{f_*} & \pi(Y,fq) \end{array}$$

form a consistent diagram and the vertical mappings are isomorphisms onto (cf (3.1)). Thus, for example, if either one of the homomorphisms f_* is one-one or onto, so is the other.

As we have indicated in the introduction to this chapter, the notion of a homomorphism induced by a continuous mapping is fundamental to algebraic topology. The homomorphism of the fundamental group induced by a continuous mapping provides the bridge from topology to algebra in knot theory.

The following important theorem shows how the topological properties of the function f are reflected in the homomorphism f_*.

(4.7) THEOREM. *If $f: X \to Y$ is a homeomorphism of X onto Y, the induced homomorphism $f_*: \pi(X,p) \to \pi(Y,fp)$ is an isomorphism onto for any basepoint p in X.*

The proof is a simple exercise using the properties formulated in (4.5). The functions

$$X \xrightarrow{f} Y \xrightarrow{f^{-1}} X$$

induce homomorphisms

$$\pi(X,p) \xrightarrow{f_*} \pi(Y,fp) \xrightarrow{(f^{-1})_*} \pi(X,p).$$

But the compositions $f^{-1}f$ and ff^{-1} are identity maps. Consequently, so are $(f^{-1}f)_* = f^{-1}{}_*f_*$ and $(ff^{-1})_* = f_*f^{-1}{}_*$. It follows from this fact that f_* is an isomorphism onto, which finishes the proof.

Thus, if pathwise connected topological spaces X and Y are homeomorphic, their fundamental groups are isomorphic. It was observed in consideration of Figure 11 that certain of the topological characteristics of X were reflected in the equivalence classes of loops of X. Theorem (4.7) is a precise formulation of this observation.

Suppose we are given two knots K and K' and we can show that the groups $\pi(R^3 - K)$ and $\pi(R^3 - K')$ are not isomorphic. By the fundamental Theorem (4.7), it then follows that $R^3 - K$ and $R^3 - K'$ are not topologically equivalent spaces. But if K and K' were equivalent knots, there would exist a homeomorphism of R^3 onto R^3 transforming K onto K'. This mapping restricted to $R^3 - K$ would give a homeomorphism of $R^3 - K$ onto $R^3 - K'$. We may conclude therefore that K and K' are knots of different type. It is by this method that many knots can be distinguished from one another.

5. Fundamental group of the circle. With a little experience it is frequently rather easy to guess correctly what the fundamental group of a not-too-complicated topological space is. Justifying one's guess with a proof, however, is likely to require topological techniques beyond a simple knowledge of the definition of the fundamental group. Chapter V is devoted to a discussion of some of these methods.

An exception to the foregoing remarks is the calculation of the fundamental group of any convex set. A subset of an n-dimensional vector space over the real or complex numbers is called *convex* if any two of its points can be joined by a straight line segment which is contained in the subset. Any p-based loop in such a set is equivalent to a constant path. To prove this we have only to set

$$h_s(t) = sp + (1 - s)a(t), \quad 0 \leq t \leq \| a \|, \quad 0 \leq s \leq 1.$$

The deformation is linear along the straight line joining p and $a(t)$. A pathwise connected space is said to be *simply-connected* if its fundamental group is trivial. As a result we have

(5.1) *Every convex set is simply-connected.*

We next consider the problem of determining the fundamental group of the circle. Our solution is motivated by the theory of covering spaces,[3] one of the topological techniques referred to in the first paragraph of this section. Let the field of real numbers be denoted by R and the subring of integers by J. We denote the additive subgroup consisting of all integers which are a multiple of 3 by $3J$. The circle, whose fundamental group we propose to calculate, may be regarded as the factor group $R/3J$ with the *identification topology*, i.e., the largest topology such that the canonical homomorphism ϕ: $R \to R/3J$ is a continuous mapping. A good way to picture the situation is to regard $R/3J$ as a circle of circumference 3 mounted like a wheel on the real line R so that it may roll freely back and forth without skidding. The possible points of tangency determine the many-one correspondence ϕ (cf. Figure 13). Incidentally, the reason for choosing $R/3J$ for our circle instead of R/J (or R/xJ for some other x) is one of convenience and will become apparent as we proceed.

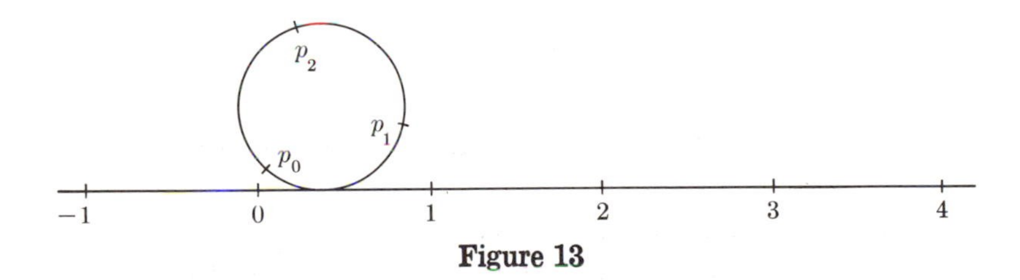

Figure 13

(5.2) *The image under ϕ of any open subset of R is an open subset of $R/3J$.*

Proof. For any subset B of $R/3J$, B is open if and only if $\phi^{-1}(B)$ is open. Furthermore, for any subset X of R,

$$\phi^{-1}\phi(X) = \bigcup_{n \in J} (3n + X),$$

where $3n + X$ is the set of all real numbers $3n + x$ with $x \in X$. Since translation along R by a fixed amount is a homeomorphism, and the union of any collection of open sets is open, our contention follows.

The mapping ϕ restricted to any interval of R of length less than 3 is one-one and, by virtue of (5.2), is therefore also a homeomorphism on that interval.

[3] H. Seifert and W. Threlfall, *Lehrbuch der Topologie*, (Teubner, Leipzig and Berlin, 1934), Ch. VIII. Reprinted by the Chelsea Publishing Co., New York, 1954.

Thus, ϕ is locally a homeomorphism. For any integer n, we define the set C_n to be the image under ϕ of the open interval $(n-1, n+1)$. It follows from (5.2) that each C_n is open and from the above remarks that the mapping

$$\phi_n\colon (n-1, n+1) \to C_n$$

defined by setting $\phi_n(x) = \phi(x)$, $n-1 < x < n+1$, is a homeomorphism. The sets C_n form an open cover of the circle. However, this cover consists of only three distinct sets because, as is easily shown,

$$C_n = C_m \text{ if and only if } \phi(n) = \phi(m).$$

Moreover, the three points

$$p_0 = \phi(0), \qquad p_1 = \phi(1), \qquad p_2 = \phi(2),$$

are the only distinct members of the image set ϕJ. Geometrically, of course, p_0, p_1, p_2, are three equally spaced points on the circle (cf. Figure 13), C_0 is the open, connected arc of length 2 running from p_1 to p_2 and containing p_0, etc.

We next define a sequence of continuous functions ψ_n by composing ϕ_n^{-1} with the inclusion mapping into R.

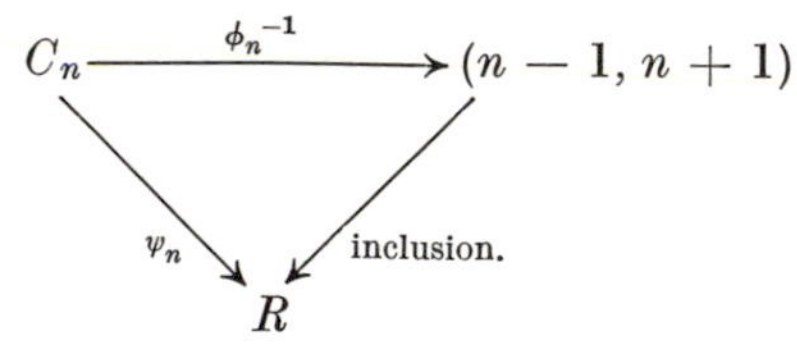

The important properties of these mappings are summarized in

(5.3)

(i) $\phi\psi_n(p) = p$, *whenever* $\psi_n(p)$ *is defined.*

(ii) *If* $\psi_n(p)$ *and* $\psi_m(p)$ *are defined, then they are equal if and only if* $|n - m| < 2$.

(iii) *For any real* x *and integer* n, *if* $\phi(x) \in C_n$, *there is exactly one integer* $m \equiv n \pmod 3$ *such that*

$$\psi_m\phi(x) = x.$$

Proof. (i) is immediate, so we pass to (ii). In one direction the result is obvious since, if $|n - m| \geq 2$, the images of ψ_n and ψ_m are disjoint. The other direction may be proved by proving that if $p \in C_n \cap C_{n+1}$, then $\psi_n(p) = \psi_{n+1}(p)$. By (i), we have that

$$p = \phi\psi_n(p) = \phi\psi_{n+1}(p).$$

Hence $\psi_n(p) = \psi_{n+1}(p) + 3r$ for some integer r. Since $\psi_n(p)$ and $\psi_{n+1}(p) \in (n-1, n+2)$, it follows that $r = 0$, and the proof of (ii) is complete. In proving (iii), we observe first of all that uniqueness is an immediate consequence of (ii). Existence is proved as follows: If $\phi(x) \in C_n$, then $\phi(x) = \phi(y)$ for some $y \in (n-1, n+1)$. Then, $x = y + 3r$, for some integer r, and

$x \in (3r + n - 1, 3r + n + 1)$. Hence,

$$\psi_{3r+n}\phi(x) = x,$$

and we may set $m = n + 3r$. This completes the proof.

Consider two arbitrary non-negative real numbers σ and τ and the rectangle E consisting of all pairs (s,t) such that $0 \leq s \leq \sigma$ and $0 \leq t \leq \tau$. The major step in our derivation of the fundamental group of the circle is the following:

(5.4) *For any continuous mapping h: $E \to R/3J$ and real number $x \in R$ such that $\phi(x) = h(0,0)$, there exists one and only one continuous function $\bar{h}$: $E \to R$ such that $\bar{h}(0,0) = x$ and $h = \phi\bar{h}$.*

Proof of uniqueness. Suppose there are two continuous mappings $\bar{h}$ and $\bar{h}'$ satisfying $h = \phi\bar{h} = \phi\bar{h}'$ and $x = \bar{h}(0,0) = \bar{h}'(0,0)$. Let E_0 be the set of all points $(s,t) \in E$ for which $\bar{h}(s,t) = \bar{h}'(s,t)$. Since R is a Hausdorff space, it is clear that E_0 is a closed subset of E. Moreover, E_0 contains the point $(0,0)$ and is therefore nonvoid. We contend that E_0 is also open. Suppose $\bar{h}(s_0,t_0) = \bar{h}'(s_0,t_0) = x_0$. For some integer n, $x_0 \in (n - 1, n + 1)$ and consequently there exist open subsets U and U' of E containing (s_0,t_0) such that both $\bar{h}U$ and $\bar{h}'U'$ are subsets of $(n - 1, n + 1)$. Then, for any $(s,t) \in U \cap U'$,

$$\bar{h}(s,t), \bar{h}'(s,t) \in (n - 1, n + 1),$$

and

$$\phi_n\bar{h}(s,t) = h(s,t) = \phi_n\bar{h}'(s,t).$$

Since ϕ_n is a homeomorphism, $\bar{h}(s,t) = \bar{h}'(s,t)$, and our contention is proved. Since E is connected, it follows that $E_0 = E$, and the proof of uniqueness is complete.

Proof of existence. We first assume that the rectangle E is not degenerate, i.e., that both σ and τ are positive. Consider a subdivision

$$0 = s_0 < s_1 < \cdots < s_k = \sigma,$$
$$0 = t_0 < t_1 < \cdots < t_l = \tau$$

which is so fine that each elementary rectangle E_{ij} defined by the inequalities $s_{i-1} \leq s \leq s_i$ and $t_{j-1} \leq t \leq t_j$ is contained in one of the open sets $h^{-1}C_n$. (Were no such subdivision to exist, there would have to be a point of E contained in rectangles of arbitrarily small diameter, no one of which would lie in any set of $h^{-1}C_n$, and this would quickly lead to a contradiction.[4]) Then there exists a function $\nu(i,j) = 0, 1, 2$, such that

$$h(E_{ij}) \subset C_{\nu(i,j)} \qquad \begin{array}{l} i = 1, \cdots, k, \\ j = 1, \cdots, l. \end{array}$$

[4] M. H. A. Newman, *Elements of the Topology of Plane Sets of Points*, Second Edition, (Cambridge University Press, Cambridge, 1951), p. 46.

The function $\bar{h}$ is constructed bit by bit by defining its values on a single elementary rectangle at a time. Starting with E_{11}, we have

$$\phi(x) = h(0,0) \in C_{\nu(1,1)}.$$

Hence, by (5.3) (iii), there exists a unique integer $\mu(1,1) \equiv \nu(1,1) \pmod 3$ such that

$$\psi_{\mu(1,1)} h(0,0) = x.$$

We define $\bar{h}(s,t) = \psi_{\mu(1,1)} h(s,t)$, for any $(s,t) \in E_{11}$. We next assume that $\bar{h}$ is extended by adjoining elementary rectangles to its domain in some order subject only to the restriction that $E_{i,j-1}$ and $E_{i-1,j}$ are always adjoined before E_{ij}. To extend to E_{ij}, we use (5.3) (iii) again to obtain a unique integer $\mu(i,j) \equiv \nu(i,j) \pmod 3$ such that

$$\psi_{\mu(i,j)} h(s_{i-1}, t_{j-1}) = \bar{h}(s_{i-1}, t_{j-1}),$$

and define $\bar{h}(s,t) = \psi_{\mu(i,j)} h(s,t)$, for any $(s,t) \in E_{ij}$. That the extension fits continuously with the previous construction is proved by using the point $h(s_{i-1}, t_{j-1})$ and (5.3) (ii) in one direction in order to conclude that

$$| \mu(i-1,j) - \mu(i,j) | < 2,$$
$$| \mu(i,j-1) - \mu(i,j) | < 2.$$

Then, from (5.3) (ii) in the other direction, it follows that $\bar{h}$ is well-defined on the left and bottom edges of E_{ij}. In this manner $\bar{h}$ is extended to all of E. The proof for a degenerate E is a corollary of the result for a nondegenerate rectangle. For example, if $\sigma = 0$ and $\tau > 0$, we choose an arbitrary $\sigma' > 0$ and define

$$h'(s,t) = h(0,t), \quad 0 \leq t \leq \tau, \quad 0 \leq s \leq \sigma'.$$

The existence of $\bar{h}'$ is assured and we set

$$\bar{h}(0,t) = \bar{h}'(0,t), \quad 0 \leq t \leq \tau.$$

The proof of (5.4) is complete.

Consider a loop a in the circle based at $p_0 = \phi(0)$. Its domain $[0, \| a \|]$ is a degenerate rectangle. It follows from (5.4) that there exists one and only one path $\bar{a}$ covering a and starting at 0, i.e., $a = \phi\bar{a}$ and $\bar{a}(0) = 0$. Since $\phi\bar{a}(\| a \|) = \phi(0)$, we know that $\bar{a}(\| a \|) = 3r$ for a uniquely determined integer $r = r_a$, which we call the *winding number* of a. Geometrically, r_a is the algebraic number of times the loop a wraps around the circle.

(5.5) $$r_{a \cdot b} = r_a + r_b.$$

Proof. Let $\bar{a}$ and $\bar{b}$ be the paths starting at 0 and covering a and b, respectively. The function $\bar{c}$ defined by

$$\bar{c}(t) = \begin{cases} \bar{a}(t), & 0 \leq t \leq \| a \|, \\ \bar{b}(t - \| a \|) + 3r_a, & \| a \| \leq t \leq \| a \| + \| b \|, \end{cases}$$

is obviously a path with initial point 0 and covering the product $a \cdot b$. Since there is only one such path, it follows immediately that

$$\begin{aligned} 3r_{a\cdot b} &= \bar{c}(\| a \| + \| b \|) = \bar{b}(\| b \|) + 3r_a \\ &= 3(r_b + r_a). \end{aligned}$$

(5.6) *Loops with equal winding numbers are equivalent.*

Proof. This result is an immediate consequence of the obvious fact that all paths in R with the same initial point and the same terminal point are equivalent. Let a and b be two p_0-based loops in the circle whose winding numbers are equal and defined by paths $\bar{a}$ and $\bar{b}$ in R. The images $h_s = \phi\bar{h}_s$, $0 \leq s \leq 1$, of any fixed-endpoint family $\{\bar{h}_s\}$ which exhibits the equivalence of $\bar{a}$ and $\bar{b}$ constitute a continuous family which proves that a is equivalent to b.

(5.7) *Equivalent loops have equal winding numbers.*

Proof. It is here that the full force of (5.4) is used. We consider a continuous family of p_0-based loops h_s, $0 \leq s \leq 1$, in the circle. Let τ be an upper bound of the set of real numbers $\| h_s \|$, $0 \leq s \leq 1$. We define a continuous function h by

$$h(s,t) = \begin{cases} h_s(t), & 0 \leq s \leq 1 \text{ and } 0 \leq t \leq \| h_s \|, \\ h_s(\| h_s \|), & 0 \leq s \leq 1 \text{ and } \| h_s \| \leq t \leq \tau. \end{cases}$$

Then, where $\bar{h}$ is the unique function covering h, i.e., $\phi\bar{h} = h$ and $\bar{h}(0,0) = 0$, we have

$$\phi\bar{h}(s, \| h_s \|) = h_s(\| h_s \|) = p_0 = \phi(0).$$

Hence, the set of image points $\bar{h}(s, \| h_s \|)$, $0 \leq s \leq 1$, is contained in the discrete set $3J$. *But a continuous function which maps a connected set into a discrete set must be constant on that set.* With this fact and the uniqueness property of covering paths we have

$$3r_{h_0} = \bar{h}(0, \| h_0 \|) = \bar{h}(1, \| h_1 \|) = 3r_{h_1},$$

and the proof is complete.

By virtue of (5.7), we may unambiguously associate to any element of $\pi(R/3J,p_0)$ the winding number of any representative loop. The definition of multiplication in the fundamental group and (5.5) show that this association is a homomorphism into the additive group of integers. (5.6) proves that the homomorphism is, in fact, an isomorphism. With the observation that there exists a loop whose winding number equals any given integer we complete the proof of the following theorem.

(5.8) *The fundamental group of the circle is infinite cyclic.*

EXERCISES

1. Compute the fundamental group of the union of two cubes joined at one corner and otherwise disjoint.

2. Compute the fundamental group of a five-pointed star (boundary plus interior).

3. Prove that if $\alpha,\beta \in \pi(X,p)$ and $a \in \alpha$, $b \in \beta$, then the loops a and b are freely equivalent (also called freely homotopic) if and only if α and β are conjugate in $\pi(X,p)$. (The definitions of "conjugate" and "freely homotopic" are given in the index.)

4. Show that if X is a simply connected space and f and g are paths from $p \in X$ to $q \in X$, then f and g belong to the same fixed-endpoint family.

5. Let $f: X \to Y$ be a continuous mapping, and $f_*: \pi(X,p) \to \pi(Y,fp)$ the induced homomorphism. Are the following statements true or false?

(a) If f is onto, then f_* is onto.

(b) If f is one-one, then f_* is one-one.

6. Prove that if X, Y, and $X \cap Y$ are nonvoid, open, pathwise connected subsets of $X \cup Y$ and if X and Y are simply-connected, then $X \cup Y$ is also simply-connected.

7. Let the definition of continuous family of paths be weakened by requiring that the function h be continuous in each variable separately instead of continuous in both simultaneously. Define the "not so fundamental group" $\pi(X,p)$ by using this weaker definition of equivalence. Show that the "not so fundamental group" of a circle is the trivial group.

CHAPTER III

The Free Groups

Introduction. In many applications of group theory, and specifically in our subsequent analysis of the fundamental groups of the complementary spaces of knots, the groups are described by "defining relations," or, as we are going to say later, are "presented". We have here another (and completely different) analogy with analytic geometry. In analytic geometry a coördinate system is selected, and the geometric configuration to be studied is defined by a set of one or more equations. In the theory of group presentations the rôle that is played in analytic geometry by a coördinate system is played by a free group. Therefore, the study of group presentations must begin with a careful description of the free groups.

1. The free group $F[\mathscr{A}]$. Let us assume that we have been given a set $\mathscr{A}$ of cardinality $\mathfrak{a}$. The elements a,b,c of $\mathscr{A}$ may be abstract symbols or they may be objects derived from some other mathematical context. We shall call $\mathscr{A}$ an *alphabet* and its members *letters*. By a *syllable* we mean a symbol a^n where a is a letter of the alphabet $\mathscr{A}$ and the exponent n is an integer. By a *word* we mean a finite ordered sequence of syllables. For example $b^{-3}a^0a^1c^2c^2a^0c^1$ is a seven-syllable word. In a word the syllables are written one after another in the form of a formal product. Every syllable is itself a word—a one-syllable word. A syllable may be repeated or followed by another syllable formed from the same letter. There is a unique word that has no syllables; it is called the *empty word*, and we denote it by the symbol 1. The syllables in a word are to be counted from the left. Thus in the example above a^1 is the third syllable. For brevity a syllable of the form a^1 is usually written simply as a.

In the set $W(\mathscr{A})$ of all words formed from the alphabet $\mathscr{A}$ there is defined a natural multiplication: the *product* of two words is formed simply by writing one after the other. The number of syllables in this product is the sum of the number of syllables in each word. It is obvious that this multiplication is associative and that the empty word 1 is both a left and a right identity. Thus $W(\mathscr{A})$ is *a semi-group*.

However $W(\mathscr{A})$ is by no means a group. In fact, the only element of $W(\mathscr{A})$ that has an inverse is 1. In order to form a group we collect the words together into equivalence classes, using a process analogous to that by which the fundamental group is obtained from the semi-group of p-based loops.

If a word u is of the form $w_1a^0w_2$, where w_1 and w_2 are words, we say that the word $v = w_1w_2$ is obtained from u by an *elementary contraction of type* I or that u is obtained from v by an *elementary expansion of type* I. If a^0 is the nth syllable of the word u, the contraction *occurs at the* nth *syllable*.

If a word u is of the form $w_1a^pa^qw_2$, where w_1 and w_2 are words, we say that the word $v = w_1a^{p+q}w_2$ is obtained from u by an *elementary contraction of type* II or that u is obtained from v by an *elementary expansion of type* II. The contraction *occurs at the* nth *syllable* if a^q is the nth syllable.

Words u and v are called *equivalent* (the relation is written $u \sim v$) if one can be obtained from the other by a finite sequence of elementary expansions and contractions. It is trivial that this is actually an equivalence relation; $W(\mathscr{A})$ is thus partitioned into equivalence classes. As before, we denote by $[u]$ the equivalence class represented by the word u. Thus, $[u] = [v]$ means the same as $u \sim v$. We denote by $F[\mathscr{A}]$ the set of equivalence classes of words.

It is easy to verify that if v' is obtained from v by an elementary contraction, then uv' is also obtained from uv by an elementary contraction, and that if u' is obtained from u by an elementary contraction, then $u'v'$ is also obtained from uv' by an elementary contraction. From this it is easy to deduce that if $u \sim u'$ and $v \sim v'$, then $uv \sim u'v'$. In other words $F[\mathscr{A}]$ inherits the multiplication of $W(\mathscr{A})$, and the inherited multiplication is defined as follows: $[u][v] = [uv]$. The associativity of the multiplication in $F[\mathscr{A}]$ follows immediately from the associativity of the multiplication in $W(\mathscr{A})$. The equivalence class $[1]$ is both a left and right identity. Thus $F[\mathscr{A}]$ inherits from $W(\mathscr{A})$ its semi-group structure. However in $F[\mathscr{A}]$ every element also has an inverse: the inverse $[u]^{-1}$ of the class $[u]$ is represented by the word $\bar{u}$ that is obtained from u by reversing the order of its syllables and changing the sign of the exponent of each syllable. For example, if $u = b^{-3}a^0a^1c^2c^2a^0c^1$, then $\bar{u} = c^{-1}a^0c^{-2}c^{-2}a^{-1}a^0b^3$. This shows that the semi-group $F[\mathscr{A}]$ is actually a group; it is called *the free group on the alphabet* $\mathscr{A}$. Note that we allow the empty alphabet; the resulting free group is trivial. The free group on an alphabet of just one letter is an infinite cyclic group. The abstract definition of a free group will be given in the third section, and it will be shown that the group $F[\mathscr{A}]$ is, in fact, free according to this definition. The name "free group on the alphabet $\mathscr{A}$" anticipates these developments.

2. Reduced words. It is important to be able to decide whether or not two given words u and v in $W(\mathscr{A})$ are equivalent. Of course, if one tries to transform u into v by elementary expansions and contractions and succeeds, then that is all there is to it, but if one fails, the question of equivalence remains unanswered. What is wanted is a procedure, or algorithm, for making this decision. The problem of finding such a uniform procedure is usually called the *word problem* for the free groups $F[\mathscr{A}]$. A solution to the problem is presented in the remainder of this section.

A word w is called *reduced* if it is not possible to apply any elementary contraction to it, i.e., if no syllable of w has exponent 0 and no two consecutive syllables are on the same letter. It is obvious, since elementary contraction always reduces the number of syllables, that each equivalence class of words contains at least one word that is reduced. We propose now to show that there is only one.

For any word w, the word $\rho(w)$ is defined as follows: If w is reduced, then $\rho(w) = w$. If w is not reduced, then $\rho(w)$ is the word obtained from w by an elementary contraction at the first possible syllable of w, i.e., the word $\rho(w)$ is obtained from w by an elementary contraction at the jth syllable of w, where no elementary contraction is possible at the kth syllable for any $k < j$. Note that it may be possible to apply an elementary contraction of both types at the jth syllable of w. However, this situation causes no ambiguity, for w must then be of the form $w = ua^p a^0 v$ where ua^p is a reduced word containing $j - 1$ syllables, and so either type of reduction yields $\rho(w) = ua^p v$. Clearly,

(2.1) *w is reduced if and only if $\rho(w) = w$.*

(2.2) *If u is not reduced, then $\rho(uv) = \rho(u)v$.*

The *standard reduction* of a word w is defined to be the sequence

$$w = \rho^0(w), \quad \rho(w), \quad \rho^2(w), \cdots.$$

If a word is not reduced, an application of ρ reduces the number of syllables by 1. Hence, in the standard reduction of any word w there exists a smallest nonnegative integer $r = r(w)$ such that $\rho^r(w) = \rho^{r+1}(w)$. This number r is the *reduction length* of w, and we define $w_* = \rho^r(w)$. Note that $\rho(w_*) = w_*$ and therefore w_* is a reduced word. In addition, the standard reduction becomes constant, i.e.,

$$\rho^i(w) = w_*, \quad \text{for every } i \geq r(w).$$

Since $\rho(w) \sim w$, we conclude that

(2.3) *w_* is reduced and $w \sim w_*$.*

Moreover,

(2.4) *w is reduced if and only if $w = w_*$.*

The central proposition in our solution of the word problem is

(2.5) *$u \sim v$ if and only if $u_* = v_*$.*

Proof. If $u_* = v_*$, then we have

$$u \sim u_* = v_* \sim v,$$

and so $u \sim v$. In proving the converse, we may assume that v is obtained from u by an elementary contraction.

Case I. $u = wa^0w'$ and $v = ww'$.

Let k equal the reduction length of w. We contend that

$$\rho^{k+1}(u) = \rho^k(v).$$

The proof is by induction on k. First, suppose that $k = 0$, i.e., that w is reduced. Then,

$$\rho(u) = ww' = v.$$

Next, assume that $k > 0$. By (2.2), we have

$$\rho(u) = \rho(w)a^0w', \qquad \rho(v) = \rho(w)w'.$$

The reduction length of $\rho(w)$ is $k - 1$. So the inductive hypothesis yields

$$\rho^k\rho(u) = \rho^{k-1}\rho(v),$$

and the contention is proved. Since the words in the standard reductions of u and v are eventually the same, it follows that $u_* = v_*$.

Case II. $u = wa^pa^qw'$ and $v = wa^{p+q}w'$.
Again, let k equal the reduction length of w. We contend that

$$\rho^{k+2}(u) = \rho^{k+1}(v).$$

The proof is by induction on k. First, assume that $k = 0$, i.e., that w is reduced. We consider two possibilities.

(a) The last syllable of w is not on the letter a. Then,

$$\rho(u) = wa^{p+q}w' = v,$$
$$\rho^2(u) = \rho(v).$$

(b) $w = w''a^r$. Then w'' is reduced and the last syllable of w'' is not on a. Hence,

$$\rho(u) = w''a^{r+p}a^qw',$$
$$\rho^2(u) = w''a^{r+p+q}w' = \rho(v),$$
$$\rho^2(u) = \rho(v).$$

Next, suppose that $k > 0$. By (2.2), we have

$$\rho(u) = \rho(w)a^pa^qw', \qquad \rho(v) = \rho(w)a^{p+q}w'.$$

The reduction length of $\rho(w)$ is $k - 1$. Hence, the hypothesis of induction gives

$$\rho^{k+1}\rho(u) = \rho^k\rho(v),$$

which is

$$\rho^{k+2}(u) = \rho^{k+1}(v).$$

Thus, the contention is proved. As in Case 1, we conclude that $u_* = v_*$. This completes the proof.

It follows directly from the preceding three propositions that

(2.6) *Each equivalence class of words contains one and only one reduced word. Furthermore, any sequence of elementary contractions of u must lead to the same reduced word u_*.*

Thus we have a finite algorithm for determining whether or not u and v represent the same element of $F[\mathscr{A}]$; one has only to find u_* and v_* and compare them syllable by syllable.

3. Free groups. Let G be an arbitrary group, and consider a subset E of G. The collection of subgroups of G that contain E is not vacuous since the improper subgroup G is a member of it. It is easily verified that the intersection of this collection is itself a subgroup that contains E; it is called *the subgroup generated by* E. If $E \neq \varnothing$, then the subgroup generated by E consists of all elements of G of the form $g_1^{n_1}g_2^{n_2}\cdots g_l^{n_l}$, where $g_1, g_2, \cdots, g_l \in E$ and $n_1, n_2, \cdots, n_l$ are integers. On the other hand, the subgroup of G generated by the empty set is trivial. If the subgroup generated by E is G itself, E is called a *generating set of elements* of G.

In the group $F[\mathscr{A}]$ each element can be written (in many ways) as a product of integral powers of $[a]$, $[b]$, $[c]$, $\cdots$. For example, $[a^2b^3c^{-2}] = [a]^2[b]^3[c]^{-2}$. Thus the elements $[a]$, $[b]$, $[c]$, $\cdots$ constitute a generating set of elements of $F[\mathscr{A}]$. We denote this generating set by $[\mathscr{A}]$.

Let us call a generating set E of elements of a group G a *free basis* if, given any group H, any function $\phi\colon E \to H$ can be extended to a homomorphism of G into H. (Since E generates G, such an extension is necessarily unique.) A group that has a free basis will be called *free*. The simplest free group is the trivial group 1; the empty set $E = \varnothing$ is a free basis of it.

(3.1) *A group is free if and only if it is isomorphic to* $F[\mathscr{A}]$ *for some* $\mathscr{A}$.

Proof. The group $F[\mathscr{A}]$ is free because $[\mathscr{A}]$ is a free basis of it. To show this, consider a function $\phi\colon [\mathscr{A}] \to H$. Denote by ϕ' the induced mapping of $\mathscr{A}$ into H. Extend ϕ' to a homomorphism into H of the semi-group $W(\mathscr{A})$ of words by defining

$$\phi'(a^m b^n \cdots) = (\phi'(a))^m(\phi'(b))^n \cdots,$$

and observe that if $u \sim v$ then $\phi'(u) = \phi'(v)$. It follows that ϕ' induces a homomorphism of $F[\mathscr{A}]$ into H. This homomorphism is clearly an extension of the function $\phi\colon [\mathscr{A}] \to H$; thus $[\mathscr{A}]$ is a free basis of $F[\mathscr{A}]$. If now G is any group that is mapped onto $F[\mathscr{A}]$ by an isomorphism λ, then $E = \lambda^{-1}[\mathscr{A}]$ is obviously a free basis of G, so that G must be a free group.

Conversely let G be a free group, and let E be a free basis of G. Let $F[\mathscr{A}]$ be the free group on an alphabet $\mathscr{A}$ whose cardinality is the same as that of E. Every element of $\mathscr{A}$ is a reduced word. It follows that if $a \neq b$, then $[a] \neq [b]$, and so there exists a natural one-one correspondence between $\mathscr{A}$ and $[\mathscr{A}]$. Hence, there exists a one-one correspondence $\kappa\colon E \to [\mathscr{A}]$. Since E is a free basis, the correspondence κ extends to a homomorphism ϕ of G into $F[\mathscr{A}]$. Since $[\mathscr{A}]$ is a free basis of $F[\mathscr{A}]$, the function $\kappa^{-1}\colon [\mathscr{A}] \to E$ extends to a homomorphism ψ of $F[\mathscr{A}]$ into G. The homomorphisms $\phi\psi\colon F[\mathscr{A}] \to F[\mathscr{A}]$ and $\psi\phi\colon G \to G$ are extensions of the respective functions $\kappa\kappa^{-1}\colon [\mathscr{A}] \to [\mathscr{A}]$ and $\kappa^{-1}\kappa\colon E \to E$. Since these functions are identities, they extend to the identity automorphisms of $F[\mathscr{A}]$ and G respectively. Since such extensions are unique, it follows that $\psi\phi$ and $\phi\psi$ are identity automorphisms. Thus ϕ maps G isomorphically onto $F[\mathscr{A}]$ and $\psi = \phi^{-1}$. This shows that G is isomorphic to $F[\mathscr{A}]$, and we are finished.

The above proof shows also that the cardinality of the free basis E of G is equal to the cardinality $\mathfrak{a}$ of the alphabet $\mathscr{A}$. Thus free groups G, G' are certainly isomorphic if they respectively have free bases E, E' of the same cardinality. It will be shown (cf. (4.2) Chapter IV) that conversely, if free groups G, G' have bases E, E' of different cardinalities, then they are not isomorphic. Granting this, it follows that to each free group G there corresponds a number n such that each free basis of G has cardinality exactly n. The cardinal number n is called the *rank* of the free group G.

(3.2) *Any group is a homomorphic image of some free group.*

This fact is of the utmost importance for the theory of group presentations; it means that, using free groups as "coördinate systems," any group can be coördinatized. Its proof is, of course, quite trivial: Let E be any set of generators of a given group G, and let $F[\mathscr{A}]$ be any free group on an alphabet $\mathscr{A}$ whose cardinality $\mathfrak{a}$ is equal to or greater than the cardinality of E. Let λ: $[\mathscr{A}] \to E$ be any function whose image is all of E. Since $[\mathscr{A}]$ is a free basis of $F[\mathscr{A}]$, the function λ extends to a homomorphism of the free group $F[\mathscr{A}]$ onto G.

EXERCISES

1. In how many ways can the word $a^{-2}bb^{-1}a^7bb^{-1}a^{-1}$ be reduced to a^4 by elementary contractions?

2. Develop a finite algorithm for determining whether or not two given words represent conjugate elements of $F[\mathscr{A}]$.

3. Develop a finite algorithm for determining whether or not a given word represents the nth power of an element of $F[\mathscr{A}]$.

4. Prove that the elements y, xyx^{-1}, x^2yx^{-2}, $\cdots$ constitute a free basis of the subgroup of $F(x,y)$ which they generate. Deduce that the free group of any given finite rank n can be mapped isomorphically into the free group of any given rank $m \geq 2$.

5. Prove that the free group of rank n cannot be generated by fewer than n elements.

6. It is known (Nielsen, Schreier, etc.[1]) that every subgroup of a free group is free. Using this fact, prove that in a free group:
   (a) There are no elements of finite order (other than the identity).
   (b) If two elements commute they are powers of a third element.
   (c) If $u^{md} = v^{nd}$, where m and n are relatively prime, then there is an element w such that $u = w^n$, $v = w^m$.
   (d) If $uvu = v$ then $u = 1$.

7. In Exercise 6 prove (a), (b), (c), (d) directly without using the Nielsen theorem.

8. Show that if u,v,u',v' are elements of a free group such that $uvu^{-1}v^{-1} = u'v'(u')^{-1}(v')^{-1} \neq 1$, then u and u' need not commute.

[1] See R. H. Fox, "Free Differential Calculus III. Subgroups," *Annals of Mathematics*, Vol. 64 (1956), p. 408.

CHAPTER IV

Presentation of Groups

Introduction. In this chapter we give a firm foundation to the concept of defining a group by generators and relations. This is an important step; for example, if one is not careful to distinguish between the elements of a group and the words that describe these elements, utter confusion is likely to ensue.

The principal problem which arises is that of recognizing when two sets of generators and relations actually present the same group. Theoretically a solution is given by the Tietze theorem. However, this leads to practical results only when coupled with some kind of systematic simplification of the groups involved. Such systematic simplification is accomplished very neatly by the so-called word subgroups, which are going to be introduced toward the end of this chapter.

1. Development of the presentation concept. The concept of an abstract group was derived from the concept of a permutation group (or substitution group as it was called), and this was, naturally, a finite group. Thus, when workers began to develop a theory of abstract groups they centered attention almost exclusively on finite groups, and so a group was usually described by exhibiting its Cayley group table. Of course, the use of a group table is not usually possible for an infinite group, nor even very practical for a finite group of large order. Furthermore the group table contains redundant information, so that it is not a very efficient device. For example, the table

	1	a	b
1	1	a	b
a	a	b	1
b	b	1	a

has nine entries, but, using the fact (obtained from the middle entry) that $b = a^2$, we can reduce the information necessary to determine the group to the statement that the elements of the group are 1, a, and a^2 and the fact that $a^3 = 1$. Thus the group in question is more efficiently depicted if we note that the element a generates the group, that the equation $a^3 = 1$ is satisfied, and that neither of the equations $a^2 = 1$ or $a = 1$ is satisfied.

This leads to the method of describing a group by giving generators and relations for it. As introduced by Dyck in 1882–3, it ran about like this: a group G is determined if there is given a set of elements $g_1, g_2, \cdots$, called *generators*, that generate the group, and a set of equations $f_1(g_1, g_2, \cdots) = 1$, $f_2(g_1, g_2, \cdots) = 1, \cdots$, called *defining equations* or *defining relations*, that have the property that every true relation that subsists among the elements $g_1, g_2, \cdots$ is an algebraic consequence of the given equations.

Now from a stricter point of view this procedure is somewhat vague in that the left-hand sides of the equations do not have true existence. What kind of an object is $f_i(g_1, g_2, \cdots)$? It cannot be an element of G, for if it were in G, it would have to be the identity element 1. In order to write down such equations we must postulate the existence of some realm in which $f_i(g_1, g_2, \cdots)$ has an independent existence. Clearly, the object required is just the free group. Thus we are led to the following reformation of the method of description.

Let F be a free group with free basis $x_1, x_2, \cdots$ in one-one correspondence with the generators $g_1, g_2, \cdots$ of G. Let ϕ be the homomorphism of F onto G defined by $\phi x_j = g_j$, $j = 1, 2, \cdots$. For each of the defining equations $f_i(g_1, g_2, \cdots) = 1$, set

$$r_i = f_i(x_1, x_2, \cdots), \qquad i = 1, 2, \cdots.$$

That is, r_i is the element of F obtained by replacing each occurrence of g_j, $j = 1, 2, \cdots$, in the expression $f_i(g_1, g_2, \cdots)$ by x_j. For example, if the ith equation is $g_1 g_2 g_1^{-1} g_2^{-1} = 1$, then $r_i = x_1 x_2 x_1^{-1} x_2^{-1}$. The assertion that the equation $f_i(g_1, g_2, \cdots) = 1$ holds in G is then equivalent to the statement that r_i is in the kernel of ϕ. Thus,

$$1 = \phi r_i = f_i(g_1, g_2, \cdots).$$

The elements $r_1, r_2, \cdots$ are called *relators*.

It is now easy to say exactly what it means for an equation to be an algebraic consequence of some others. Remembering that we have replaced each equation $f_i(g_1, g_2, \cdots) = 1$ by a group element r_i, we see that the following must be meant. An element f of an arbitrary group Q is called a *consequence* of a set of elements $f_1, f_2, \cdots$ in Q if every homomorphism ψ of Q into any group H that maps each of the elements $f_1, f_2, \cdots$ into 1 also maps the element f into 1. Since every homomorphism of Q determines a normal subgroup, i.e., the kernel of the homomorphism, and conversely, since every normal subgroup of Q determines a homomorphism of which it is the kernel; the definition can be rephrased as follows: An element f of Q is a *consequence* of elements $f_1, f_2, \cdots$ if f is contained in every normal subgroup of Q that contains all the elements $f_1, f_2, \cdots$. Let us call the set of all consequences of $f_1, f_2, \cdots$ *the consequence* of $f_1, f_2, \cdots$. Then what we have found is that the consequence is the intersection of all the normal subgroups of Q which contain all the elements $f_1, f_2, \cdots$. Since the intersection of any collection of normal subgroups is itself a normal subgroup, we can also say that the consequence of $f_1, f_2, \cdots$ is the smallest normal subgroup of Q which contains all the elements $f_1, f_2, \cdots$.

We can determine the consequences of $f_1, f_2, \cdots$ even more explicitly.

Observe that any product of transforms of powers of these elements; i.e., any element k of the form

$$\prod_{j=1}^{l} h_j f_{i(j)}^{n(j)} h_j^{-1}$$

is mapped into 1 by any homomorphism that maps each of $f_1, f_2, \cdots$ into 1. Hence every such element k is a consequence of $f_1, f_2, \cdots$. It is easy to see that the set of all such elements k constitutes a normal subgroup K of Q; thus K is contained in the consequence of $f_1, f_2, \cdots$. On the other hand, K is a normal subgroup of Q which contains all the elements $f_1, f_2, \cdots$, and so it is one of the normal subgroups belonging to the collection whose intersection is the consequence of $f_1, f_2, \cdots$. It follows that the consequence of $f_1, f_2, \cdots$ is just K. We have therefore shown that an element of Q is a consequence of $f_1, f_2, \cdots$ if and only if it is of the form

$$\prod_{j=1}^{l} h_j f_{i(j)}^{n(j)} h_j^{-1}.$$

We shall have occasion to use the following theorem.

(1.1) *Let $g_1, g_2, \cdots$ be a set of elements of a group G, and let ϕ be a homomorphism of G onto a group H. Then ϕ maps the consequence of $g_1, g_2, \cdots$ onto the consequence of the set $\phi g_1, \phi g_2, \cdots$ of elements of H.*

Proof. Denote the consequence of $g_1, g_2, \cdots$ by K_G and the consequence of $h_1 = \phi g_1, h_2 = \phi g_2, \cdots$ by K_H. Since ϕK_G contains all the elements $h_1, h_2, \cdots$ and is normal, ϕK_G must contain K_H. To prove the reverse inclusion, consider any element h of ϕK_G and select an element $g \in K_G$ such that $\phi g = h$. If ψ is any homomorphism of H that maps each of the elements $h_1, h_2, \cdots$ into 1, then $\psi\phi$ must map each of the elements $g_1, g_2, \cdots$ into 1. Since $g \in K_G$, we must have $\psi\phi g = 1$, that is to say, $\psi h = 1$. Since h is mapped into 1 by every such homomorphism, h must belong to K_H. This shows that ϕK_G is contained in K_H, and therefore $\phi K_G = K_H$. This completes the proof.

Returning now to the homomorphism $F \xrightarrow{\phi} G$, we denote by R the consequence of the relators $r_1, r_2, \cdots$. The assertion that the equations

$$f_i(g_1, g_2, \cdots) = 1, \quad i = 1, 2, \cdots,$$

constitute a defining set of relations for G from which all others can be derived is simply the assertion that R equals the kernel of ϕ. In this case the group G is determined by the free basis $x_1, x_2, \cdots$ and the elements $r_1, r_2, \cdots$ because G is isomorphic to the factor group F/R.

2. Presentations and presentation types. The following definitions formalize the ideas of the preceding section. Let F be a free group with a free basis E that is supposed to be large enough to include an inexhaustible supply of basic elements.

The set E is called the *underlying set of generators*. A *group presentation*, denoted by $(\mathbf{x} : \mathbf{r})$, is an object that consists of a subset $\mathbf{x}$ of the underlying set of generators and a subset $\mathbf{r}$ of the subgroup $F(\mathbf{x})$ generated in F by $\mathbf{x}$. Notice that $F(\mathbf{x})$ is isomorphic to the free group $F[\mathbf{x}]$ on the alphabet $\mathbf{x}$. It is important to observe that $F(\mathbf{x})$ is itself a free group and $\mathbf{x}$ is a free basis of it; this follows directly from the definition of free basis (without appealing to the deep Nielsen-Schreier theorem that asserts that *any* subgroup of a free group is free). The set $\mathbf{x}$ is called the set of *generators* of the presentation and the set $\mathbf{r}$ is called the set of *relators* of the presentation. The *group of*, or *defined by, a presentation* $(\mathbf{x} : \mathbf{r})$ is the factor group $|\mathbf{x} : \mathbf{r}| = F(\mathbf{x})/R$, where R is the consequence in $F(\mathbf{x})$ of $\mathbf{r}$.

A *presentation of a group* G consists of a group presentation $(\mathbf{x} : \mathbf{r})$ and an isomorphism ι of the group $|\mathbf{x} : \mathbf{r}|$ onto G. Clearly, any homomorphism ϕ of the free group $F(\mathbf{x})$ onto a group G whose kernel is the consequence of $\mathbf{r}$ determines a presentation of G. Conversely, any presentation of G determines such a homomorphism. That is, if γ denotes the canonical homomorphism of $F(\mathbf{x})$ upon $F(\mathbf{x})/R$ in the consistent diagram, then either one of ϕ and ι determines the other uniquely. When the extra precision is desired, we write

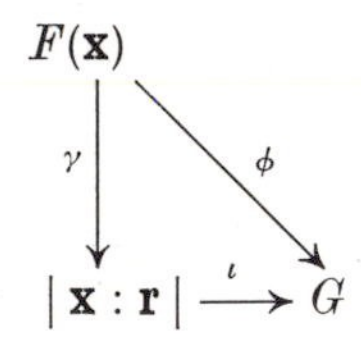

$(\mathbf{x} : \mathbf{r})_\phi$ to indicate that $(\mathbf{x} : \mathbf{r})$ is a presentation of the group G with respect to the homomorphism ϕ.

The name "presentation" was selected to describe the situation in which a group G is studied by mapping a known group (the free group $F(\mathbf{x})$) onto it because it was felt that this is in some way dual to the situation in which a group G is studied by mapping it into a known group (e.g. a group of permutations); the latter mappings are what are called "representations" of G.

Although there is no logical necessity for it we shall now reintroduce the concept of *relation*. The reason for doing this is that the manipulation of relations fits more easily into our accustomed patterns of thought than the manipulation of relators. For example, it is easy to see that if a and b commute, then the fact that $(ab)^2 = 1$ implies that $a^2b^2 = 1$, but it is not quite so easy to show that a^2b^2 is a consequence of the two relators $aba^{-1}b^{-1}$ and $(ab)^2$. (In fact, $a^2b^2 = b^{-1}(aba^{-1}b^{-1})^{-1}b \cdot b^{-1}(ab)^2b$.) It is not difficult to put the "relation" concept on a sound footing as in the following: by the formula $u = v$ is meant what would more properly be written $u \equiv v \pmod{R}$, i.e., $uv^{-1} \in R$. This is, of course, always with reference to a given presentation $(\mathbf{x} : \mathbf{r})$. On occasion we might even write $(\mathbf{x} : uv^{-1} = 1, \cdots)$ or $(\mathbf{x} : u = v, \cdots)$ meaning the same thing as $(\mathbf{x} : uv^{-1}, \cdots)$. There is no use in trying to be more precise about it, as the only advantage of the use of relations in the place of relators lies in the informality that is achieved.

The presentation notation may be used consistently even when the set of relators is empty. Thus $(\mathbf{x} \ :)$ is a presentation of the free group $F(\mathbf{x})$. Although it is unlikely that the occasion should often arise, we might even denote by $(:)$ an especially simple presentation of the trivial group.

A presentation $(\mathbf{x} \ : \ \mathbf{r})$ is *finitely generated* if $\mathbf{x}$ is finite, *finitely related* if $\mathbf{r}$ is finite. A *finite presentation* is one that is both finitely generated and finitely related. A group is said to be *finitely generated* if it has at least one finitely generated presentation, *finitely related* if it has at least one finitely related presentation, *finitely presented* if it has at least one finite presentation. Although nonfinite presentations are common enough and by no means pathological, we shall be primarily concerned with finite sets $\mathbf{x}$ and $\mathbf{r}$.

Just as the equations of a curve or of a surface take on different forms in different coördinate systems, so a group has many different presentations. For example, it may be shown that

$$| x,y \ : \ xyx = yxy \,| \approx | \, a,b \ : \ a^3 = b^2 \,|,$$

and that

$$| x,y \ : \ xy^2 = y^3x, \ yx^2 = x^3y \,| \approx | : |.$$

The problem of determining whether or not two presentations determine isomorphic groups is the *isomorphism problem.* It is not possible to give a general solution of this problem[1], but partial solutions can be found, and these are of great importance. These are usually of the nature of conditions on presentations that must be fulfilled if the groups presented are to be isomorphic. Such conditions are of importance because *they are the means of showing that certain groups are not isomorphic.* The methodology of finding such partial solutions of the isomorphism problem will now be considered.

A *mapping* f: $(\mathbf{x} \ : \ \mathbf{r}) \to (\mathbf{y} \ : \ \mathbf{s})$ *of presentations* consists of the two presentations $(\mathbf{x} \ : \ \mathbf{r})$ and $(\mathbf{y} \ : \ \mathbf{s})$ and a homomorphism f: $F(\mathbf{x}) \to F(\mathbf{y})$ which satisfies the condition that the image $f(\mathbf{r})$ of $\mathbf{r}$ under f is contained in the consequence of $\mathbf{s}$.

Every presentation mapping f: $(\mathbf{x} \ : \ \mathbf{r}) \to (\mathbf{y} \ : \ \mathbf{s})$ determines uniquely a group homomorphism f_*: $|\, \mathbf{x} \ : \ \mathbf{r} \,| \to |\, \mathbf{y} \ : \ \mathbf{s} \,|$ satisfying $f_*\gamma = \gamma f$, where the canonical homomorphisms $F(\mathbf{x}) \to |\, \mathbf{x} \ : \ \mathbf{r} \,|$ and $F(\mathbf{y}) \to |\, \mathbf{y} \ : \ \mathbf{s} \,|$ are both denoted by the symbol γ.

$$\begin{array}{ccc} F(\mathbf{x}) & \xrightarrow{f} & F(\mathbf{y}) \\ \gamma \downarrow & & \downarrow \gamma \\ |\, \mathbf{x} : \mathbf{r} \,| & \xrightarrow{f_*} & |\, \mathbf{y} : \mathbf{s} \,| \end{array}$$

[1] There are a number of similar problems which are known to have no *general* solution: deciding whether or not the group defined by a given presentation is trivial (the triviality problem), is finite, is abelian, is free, etc; deciding whether or not a given word is a consequence of a given set of words (the word problem); and many others. See M. O. Rabin, "Recursive Unsolvability of Group Theoretic Problems," *Annals of Mathematics*, Vol. 67 (1958), pp. 172–194.

Composition of presentation maps is defined in the natural way. If we are given mappings $f\colon (\mathbf{x} : \mathbf{r}) \to (\mathbf{y} : \mathbf{s})$ and $g\colon (\mathbf{y} : \mathbf{s}) \to (\mathbf{z} : \mathbf{t})$, the composition gf consists of $(\mathbf{x} : \mathbf{r})$ and $(\mathbf{z} : \mathbf{t})$ and the homomorphism $gf\colon F(\mathbf{x}) \to F(\mathbf{z})$. The associative law holds and there are identity mappings. Thus the collection of presentations and presentation mappings forms a category. Moreover, $1_* = 1$ and $(gf)_* = g_* f_*$.

Presentation mappings $f_1, f_2\colon (\mathbf{x} : \mathbf{r}) \to (\mathbf{y} : \mathbf{s})$ are *homotopic*, written $f_1 \simeq f_2$, if, for every x in $\mathbf{x}$, the element $f_1(x)f_2(x^{-1})$ belongs to the consequence of $\mathbf{s}$.

The condition for homotopy of presentation mappings can be restated: $\gamma f_1(u) = \gamma f_2(u)$ for every $u \in F(\mathbf{x})$. Since the definition of induced mapping gives $f_{i*}\gamma(u) = \gamma f_i(u)$, $i = 1, 2$, we have shown

(2.1) *$f_1 \simeq f_2$ if and only if $f_{1*} = f_{2*}$.*

Furthermore,

(2.2) *If $f_1 \simeq f_2$ and $g_1 \simeq g_2$, then $g_1 f_1 \simeq g_2 f_2$.*

We have seen that a presentation map f determines a homomorphism f_*. Conversely,

(2.3) *For each homomorphism $\theta\colon |\mathbf{x} : \mathbf{r}| \to |\mathbf{y} : \mathbf{s}|$, there exists a presentation map $f\colon (\mathbf{x} : \mathbf{r}) \to (\mathbf{y} : \mathbf{s})$ such that $f_* = \theta$. Furthermore, any two such presentation maps are homotopic.*

Proof. Consider the diagram

$$\begin{array}{ccc} F(\mathbf{x}) & & F(\mathbf{y}) \\ \gamma\downarrow & & \gamma\downarrow \\ |\mathbf{x} : \mathbf{r}| & \xrightarrow{\theta} & |\mathbf{y} : \mathbf{s}|. \end{array}$$

Since γ is onto, we can assign to each $x \in \mathbf{x}$ an element $f(x) \in F(\mathbf{y})$ in such a way that $\gamma f(x) = \theta\gamma(x)$. Since $F(\mathbf{x})$ is a free group with basis $\mathbf{x}$ this assignment may be extended to a homomorphism $f\colon F(\mathbf{x}) \to F(\mathbf{y})$ such that $\gamma f = \theta\gamma$. The image $f\mathbf{r}$ is contained in the consequence of $\mathbf{s}$; hence f is a presentation mapping, and $f_* = \theta$. The uniqueness of f up to homotopy follows from (2.1).

Thus the homotopy classes of presentation maps are in one-one correspondence with the homomorphisms between the groups presented. In addition, the correspondence is composition preserving.

Presentations $(\mathbf{x} : \mathbf{r})$ and $(\mathbf{y} : \mathbf{s})$ are of the same *type* if there exist mappings $(\mathbf{x} : \mathbf{r}) \underset{g}{\overset{f}{\rightleftarrows}} (\mathbf{y} : \mathbf{s})$ such that $gf \simeq 1$ and $fg \simeq 1$. The pair of mappings f, g (or either one separately) is called a *presentation* (or *homotopy*) *equivalence.*

(2.4) *Two presentations are of the same type if and only if their groups are isomorphic.*

Proof. If f,g is a presentation equivalence, then

$$g_*f_* = (gf)_* = 1_* = 1,$$
$$f_*g_* = (fg)_* = 1_* = 1;$$

hence f_* maps $|\mathbf{x} : \mathbf{r}|$ isomorphically upon $|\mathbf{y} : \mathbf{s}|$ (and $g_* = f_*^{-1}$). Conversely, if θ maps $|\mathbf{x} : \mathbf{r}|$ isomorphically upon $|\mathbf{y} : \mathbf{s}|$ and $f_* = \theta, g_* = \theta^{-1}$, then

$$(gf)_* = g_*f_* = \theta^{-1}\theta = 1 = (1)_*, \quad \text{hence } gf \simeq 1,$$
$$(fg)_* = f_*g_* = \theta\theta^{-1} = 1 = (1)_*, \quad \text{hence } fg \simeq 1.$$

3. The Tietze theorem. Among presentation equivalences special importance is attached to the *Tietze equivalences* **I**, **I**′, **II**, **II**′ which will now be considered.

Let $(\mathbf{x} : \mathbf{r})$ be any presentation and let s be any consequence of $\mathbf{r}$. Consider the presentation $(\mathbf{y} : \mathbf{s})$ made up of $\mathbf{y} = \mathbf{x}$ and $\mathbf{s} = \mathbf{r} \cup s$. In this case the consequence of $\mathbf{r}$ equals the consequence of $\mathbf{s}$. Hence $(\mathbf{x} : \mathbf{r})$, $(\mathbf{y} : \mathbf{s})$, and the identity automorphism $1 : F(\mathbf{x}) \to F(\mathbf{y})$ define a presentation mapping $\mathbf{I} : (\mathbf{x} : \mathbf{r}) \to (\mathbf{y} : \mathbf{s})$. Similarly, $(\mathbf{y} : \mathbf{s})$, $(\mathbf{x} : \mathbf{r})$, and the identity 1 define a presentation mapping $\mathbf{I}' : (\mathbf{y} : \mathbf{s}) \to (\mathbf{x} : \mathbf{r})$. The pair of mappings **I** and **I**′ is trivially a presentation equivalence.

Starting again from an arbitrary presentation $(\mathbf{x} : \mathbf{r})$ let y be any member of the underlying set of generators that is not contained in $\mathbf{x}$, and let ξ be any element of $F(\mathbf{x})$. Consider the presentation $(\mathbf{y} : \mathbf{s})$ made up of $\mathbf{y} = \mathbf{x} \cup y$ and $\mathbf{s} = \mathbf{r} \cup y\xi^{-1}$. The homomorphism $\mathbf{II}: F(\mathbf{x}) \to F(\mathbf{y})$, defined by the rule $\mathbf{II}(x) = x$ for any $x \in \mathbf{x}$, maps $\mathbf{r}$ into the consequence of $\mathbf{s}$ so that $(\mathbf{x} : \mathbf{r})$, $(\mathbf{y} : \mathbf{s})$, and $\mathbf{II}: F(\mathbf{x}) \to F(\mathbf{y})$ define a presentation map $\mathbf{II}: (\mathbf{x} : \mathbf{r}) \to (\mathbf{y} : \mathbf{s})$. Also the homomorphism $\mathbf{II}': F(\mathbf{y}) \to F(\mathbf{x})$ defined by the rule $\mathbf{II}'(x) = x$ for any $x \in \mathbf{x}$ and $\mathbf{II}'(y) = \xi$ maps $\mathbf{s}$ onto $\mathbf{r} \cup 1$ and hence into the consequence of $\mathbf{r}$. It follows that $(\mathbf{y} : \mathbf{s})$, $(\mathbf{x} : \mathbf{r})$, and $\mathbf{II}': F(\mathbf{y}) \to F(\mathbf{x})$ define a presentation map $\mathbf{II}': (\mathbf{y} : \mathbf{s}) \to (\mathbf{x} : \mathbf{r})$. The composition $\mathbf{II}'\mathbf{II}$ is the identity. Also for every $x \in \mathbf{x}$, $\mathbf{II}\,\mathbf{II}'(x) \cdot x^{-1} = 1$, and $\mathbf{II}\,\mathbf{II}'(y) \cdot y^{-1} = \mathbf{II}(\xi) \cdot y^{-1} = \xi y^{-1} = (y\xi^{-1})^{-1}$ which belongs to the consequence of $\mathbf{s}$, so that $\mathbf{II}\,\mathbf{II}' \simeq 1$. Thus the pair $\mathbf{II}, \mathbf{II}'$ is a presentation equivalence. Note that $\mathbf{II}: F(\mathbf{x}) \to F(\mathbf{y})$ is an inclusion and $\mathbf{II}': F(\mathbf{y}) \to F(\mathbf{x})$ is a retraction.[2]

Although theoretically **I** and **I**′ are completely trivial and **II** and **II**′ somewhat less so, in practice the opposite is true. Actually checking that an element, or proposed relation, is a consequence of certain others can be quite difficult. (It is a special case of the word problem, cf. footnote 1 on page 41.) The same difficulty occurs in the proof of the fundamental Tietze theorem that we are getting ready to prove. It is precisely in order to verify the use of **I** and **I**′ in that proof that the following lemma is needed.

(3.1) *Let* $\mathbf{x}$ *and* $\mathbf{y}$ *be disjoint sets of underlying basis elements, and let* θ *be a retraction of* $F(\mathbf{x} \cup \mathbf{y})$ *onto* $F(\mathbf{x})$. *Let* $(\mathbf{x} : \mathbf{r})_\phi$ *be a presentation of a group* G.

[2] A *retraction* is any mapping $f: X \to Y$ such that $Y \subset X$ and $f(p) = p$ for every $p \in Y$.

Then the kernel of the homomorphism $\phi\theta\colon F(\mathbf{x} \cup \mathbf{y}) \to G$ *is the consequence* C *of the union of* $\mathbf{r}$ *and the set of all elements* $y \cdot \theta(y)^{-1}$, $y \in \mathbf{y}$.

Proof. Clearly, $\phi\theta(r) = \phi(r) = 1$ for any $r \in \mathbf{r}$. Since θ is a retraction, $\theta^2 = \theta$, and hence $\phi\theta(y \cdot \theta(y)^{-1}) = \phi(\theta(y) \cdot \theta(y)^{-1}) = \phi(1) = 1$. Thus C is contained in the kernel of $\phi\theta$.

To prove the converse, we consider the canonical homomorphism γ of $F(\mathbf{x} \cup \mathbf{y})$ onto the factor group $F(\mathbf{x} \cup \mathbf{y})/C$ and its restriction $\gamma' = \gamma \mid F(\mathbf{x})$.

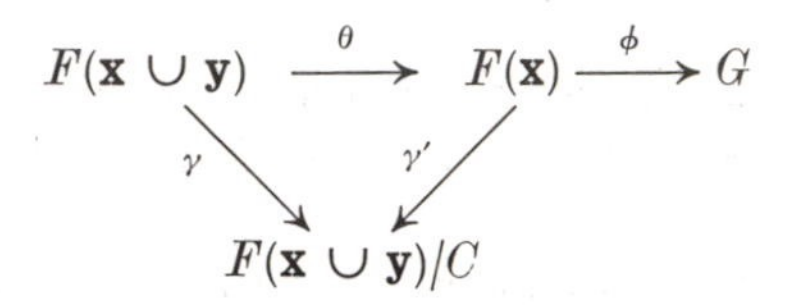

We have $\gamma'\theta(x) = \gamma'(x) = \gamma(x)$ for $x \in \mathbf{x}$. Moreover, for $y \in \mathbf{y}$, we have $\gamma(y) \cdot \gamma'\theta(y)^{-1} = \gamma(y) \cdot \gamma\theta(y)^{-1} = \gamma(y \cdot \theta(y)^{-1}) = 1$, hence $\gamma'\theta(y) = \gamma(y)$. This shows that $\gamma'\theta = \gamma$. Suppose now that $u \in F(\mathbf{x} \cup \mathbf{y})$ such that $\phi\theta(u) = 1$. Then $\gamma(u \cdot \theta(u)^{-1}) = \gamma'\theta(u \cdot \theta(u)^{-1}) = \gamma'(\theta(u) \cdot \theta(u)^{-1}) = \gamma'(1) = 1$, and so $u \cdot \theta(u)^{-1} \in C$. But, $\phi\theta(u) = 1$, so that $\theta(u)$ is in the consequence of $\mathbf{r}$ and therefore lies in C. We conclude that $u = u \cdot \theta(u)^{-1} \cdot \theta(u) \in C$.

(3.2) TIETZE THEOREM. *Suppose that* $(\mathbf{x} : \mathbf{r}) \underset{g}{\overset{f}{\rightleftarrows}} (\mathbf{y} : \mathbf{s})$ *is a presentation equivalence and that the presentations* $(\mathbf{x} : \mathbf{r})$ *and* $(\mathbf{y} : \mathbf{s})$ *are both finite. Then there exists a finite sequence* $T_1, T_1'; \cdots; T_l, T_l'$ *of Tietze equivalences such that* $f = T_1 \cdots T_l$ *and* $g = T_l' \cdots T_1'$.

Proof. We shall first prove this under the assumption that $\mathbf{x}$ and $\mathbf{y}$ are disjoint sets. We consider the following diagram

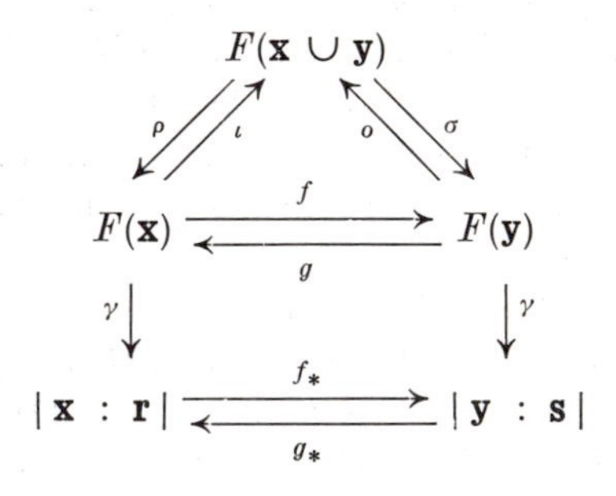

where ι and o are inclusions, and ρ and σ are retractions defined so that $\rho(y) = g(y)$ for $y \in \mathbf{y}$, and $\sigma(x) = f(x)$ for $x \in \mathbf{x}$.

It is apparent that the presentation equivalence

$$(\mathbf{x} : \mathbf{r}) \underset{\rho}{\overset{\iota}{\rightleftarrows}} (\mathbf{x} \cup \mathbf{y} : \mathbf{r} \cup \mathbf{b}),$$

where $\mathbf{b} = \{y \cdot \rho(y)^{-1}\}$, can be factored into Tietze **II** equivalences $T_1, T_1'; \cdots; T_m, T_m'$, where m is the number of elements of the set $\mathbf{y}$, so that $\iota = T_1 \cdots T_m$, $\rho = T_m' \cdots T_1'$. Similarly the presentation equivalence

$$(\mathbf{y} : \mathbf{s}) \underset{\sigma}{\overset{o}{\rightleftarrows}} (\mathbf{x} \cup \mathbf{y} : \mathbf{s} \cup \mathbf{a}),$$

where $\mathbf{a} = \{x \cdot \sigma(x)^{-1}\}$, can be factored into Tietze **II** equivalences $S_1, S_1'; \cdots; S_n, S_n'$, where n is the number of elements of the set $\mathbf{x}$, so that $o = S_1 \cdots S_n$, $\sigma = S_n' \cdots S_1'$.

Now it follows from (3.1) that the kernel of the homomorphism $\gamma\rho$ is the consequence of $\mathbf{r} \cup \mathbf{b}$. But $\gamma\rho = g_*\gamma\sigma$, and $\gamma\sigma(\mathbf{s} \cup \mathbf{a}) = 1$; hence $\mathbf{s} \cup \mathbf{a}$ is contained in the consequence of $\mathbf{r} \cup \mathbf{b}$. By the same argument $\mathbf{r} \cup \mathbf{b}$ is contained in the consequence of $\mathbf{s} \cup \mathbf{a}$. Hence the presentation equivalences

$$(\mathbf{x} \cup \mathbf{y} : \mathbf{r} \cup \mathbf{b}) \underset{\alpha'}{\overset{\alpha}{\rightleftarrows}} (\mathbf{x} \cup \mathbf{y} : \mathbf{r} \cup \mathbf{s} \cup \mathbf{a} \cup \mathbf{b}) \underset{\beta'}{\overset{\beta}{\leftrightarrows}} (\mathbf{x} \cup \mathbf{y} : \mathbf{s} \cup \mathbf{a})$$

carried by the identity automorphism of $F(\mathbf{x} \cup \mathbf{y})$ can be factored into Tietze **I** equivalences $U_1, U_1'; \cdots; U_{q+n}, U'_{q+n}$ and $V_1, V_1'; \cdots; V_{p+m}, V'_{p+m}$ respectively, where p is the number of elements of the set $\mathbf{r}$ and q is the number of elements of the set $\mathbf{s}$. Then, $\alpha = U_1 \cdots U_{q+n}$, $\alpha' = U'_{q+n} \cdots U_1'$, $\beta = V_1 \cdots V_{p+m}$, $\beta' = V'_{p+m} \cdots V_1'$, and so

$$f = \sigma\beta'\alpha\iota = S_n' \cdots S_1' V'_{p+m} \cdots V_1' U_1 \cdots U_{q+n} T_1 \cdots T_m,$$
$$g = \rho\alpha'\beta o = T_m' \cdots T_1' U'_{q+n} \cdots U_1' V_1 \cdots V_{p+m} S_1 \cdots S_n.$$

If $\mathbf{x}$ and $\mathbf{y}$ are not disjoint, we select from the underlying set of generators a subset $\mathbf{z}$ which is disjoint from $\mathbf{x} \cup \mathbf{y}$ and is in one-one correspondence with $\mathbf{x}$. This correspondence induces an isomorphism h_1 of $F(\mathbf{x})$ onto $F(\mathbf{z})$ and the inverse isomorphism $h_2 = h_1^{-1}$ of $F(\mathbf{z})$ onto $F(\mathbf{x})$. Let $\mathbf{t} = h_1(\mathbf{r})$, $k_1 = fh_2$, and $k_2 = h_1 g$, so that $f = k_1 h_1$ and $g = h_2 k_2$.

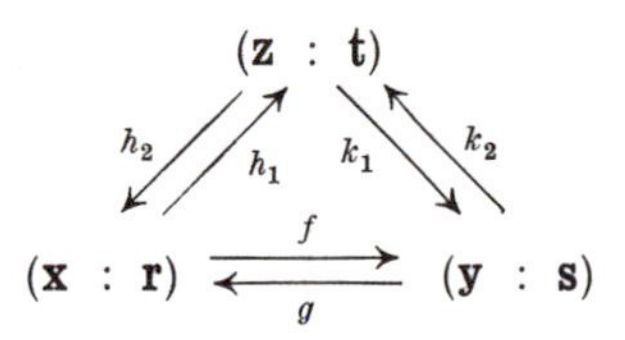

Clearly h_1, h_2 is a presentation equivalence. We claim that k_1, k_2 is also a presentation equivalence. Let the consequences of $\mathbf{r}$, $\mathbf{s}$, and $\mathbf{t}$ be denoted respectively by R, S, and T. Then $k_1(\mathbf{t}) = fh_2(\mathbf{t}) = f(\mathbf{r}) \subset S$ and $k_2(\mathbf{s}) = h_1 g(\mathbf{s}) \subset h_1(R) = T$, so that k_1 and k_2 are presentation maps; furthermore $k_2 k_1 = h_1 g f h_2 \simeq h_1 1 h_2 = 1$ and $k_1 k_2 = f h_2 h_1 g = fg \simeq 1$. Now we can apply the first part of the proof twice and we are finished.

The importance of the Tietze theorem is that it reduces the problem of showing that a given function of group presentations depends only on the group presented to one of checking that it is unaltered by the Tietze operations **I** and **II**. For example, in Chapter VII we shall show how to compute a certain sequence of so-called elementary ideals from each finite group presentation. Since presentations that differ by the two Tietze operations will be shown to give isomorphic sequences, we shall conclude that the elementary ideals are group invariants.

As an example of how Tietze equivalences are used to obtain one presentation from another, let us show that the groups $| x,y,z \ : \ xyz = yzx |$ and $| x,y,a \ : \ xa = ax |$ are isomorphic:

$$(x,y,z \ : \ xyz(yzx)^{-1})$$
$$\downarrow \ \mathbf{II}$$
$$(x,y,z,a \ : \ xyz(yzx)^{-1},\ a(yz)^{-1})$$
$$\downarrow \ \mathbf{I}$$
$$(x,y,z,a \ : \ xa(ax)^{-1},\ a(yz)^{-1},\ xyz(yzx)^{-1})$$
$$\downarrow \ \mathbf{I}'$$
$$(x,y,z,a \ : \ xa(ax)^{-1},\ a(yz)^{-1})$$
$$\downarrow \ \mathbf{I}$$
$$(x,y,z,a \ : \ xa(ax)^{-1},\ z(y^{-1}a)^{-1},\ a(yz)^{-1})$$
$$\downarrow \ \mathbf{I}'$$
$$(x,y,a,z \ : \ xa(ax)^{-1},\ z(y^{-1}a)^{-1})$$
$$\downarrow \ \mathbf{II}'$$
$$(x,y,a \ : \ xa(ax)^{-1}).$$

As another example let us show that $| x,y \ : \ xyx = yxy |$ is isomorphic to $| a,b \ : \ a^3 = b^2 |$. To see how this could be done, we begin by noticing that if we multiply both sides of $xyx = yxy$ on the left by xyx, we get $(xyx)(xyx) = (xy)(xy)(xy)$. Then we set $a = xy$ and $b = xyx$ and observe that these last two relations can be solved for x and y. This reasoning leads us to the following sequence of Tietze equivalences, which we now write in the informal style:

$$(x,y \ : \ xyx = yxy)$$
$$\downarrow \ \mathbf{II} \text{ (twice)}$$
$$(x,y,a,b \ : \ xyx = yxy,\ a = xy,\ b = xyx)$$
$$\downarrow \ \mathbf{I} \text{ (thrice)}$$
$$(x,y,a,b \ : \ xyx = yxy,\ a^3 = b^2,\ a = xy,\ b = xyx,\ x = a^{-1}b,\ y = b^{-1}a^2)$$
$$\downarrow \ \mathbf{I}' \text{ (thrice)}$$
$$(x,y,a,b \ : \ a^3 = b^2,\ x = a^{-1}b,\ y = b^{-1}a^2)$$
$$\downarrow \ \mathbf{II}' \text{ (twice)}$$
$$(a,b \ : \ a^3 = b^2).$$

4. Word subgroups and the associated homomorphisms. If one wishes to find necessary conditions in the isomorphism problem, one is almost forced to find some uniform method of simplifying groups. To accomplish this we are going to make use of the "word subgroups," as will now be explained.

To define a word subgroup we begin by selecting a subset W of some free group $F(\mathbf{x})$. (The elements of W are represented by words in the underlying set of generators, and this is the origin of the term "word subgroup.") Given any group G we consider the set $\Omega = \Omega(G)$ of all possible homomorphisms ω of $F(\mathbf{x})$ into G, and we denote by $W(G)$ the subgroup of G generated by all the elements $\omega(w)$, $w \in W$, $\omega \in \Omega$. Such a subgroup $W(G)$, which is called a *word subgroup*, is necessarily a normal subgroup since it is unaltered by any inner automorphism of G. In fact $W(G)$ has an even stronger property—it is mapped into itself by every endomorphism of G (such subgroups are called *fully normal*). For if α is any endomorphism of G and if $\omega \in \Omega$, then $\alpha\omega \in \Omega$ so that $\alpha(W(G)) \subset W(G)$.

The simplest examples of word subgroups are the *commutator subgroup* and the *power*. To obtain the commutator subgroup we select $\mathbf{x} = \{x,y\}$ and W the subset of $F(x,y)$ consisting of the single element $[x,y] = xyx^{-1}y^{-1}$. The resulting word subgroup $W(G)$ is called the *commutator subgroup* and may be denoted $[G,G]$. It is the subgroup of G generated by all elements of the form $g_1g_2g_1^{-1}g_2^{-1}$. The quotient group $G/[G,G]$ is called the *commutator quotient group* or the *abelianized group*, and the canonical homomorphism $G \to G/[G,G]$ is called the *abelianizer*. The commutator quotient group is an abelian group; abelianization just has the effect of making everything commute.

To obtain the nth power ($n \geq 0$) of G we select $\mathbf{x} = \{x\}$ and W the subset of $F(x)$ consisting of the single element x^n. The resulting word subgroup $W(G)$ is called the nth *power* of G and may be denoted G^n. It is the subgroup of G generated by all elements of the form g^n. It should be clear that $G^0 = 1$, $G^1 = G$, and that $G^m \subset G^n$ whenever m is divisible by n. Also it may be noted that $[G,G] \subset G^2$; in fact, $g_1g_2g_1^{-1}g_2^{-1} = (g_1g_2)^2 \cdot (g_2^{-1}g_1^{-1}g_2)^2 \cdot g_2^{-2}$, and this means that G/G^2 is always abelian.

If ϕ: $G_1 \to G_2$ is a group homomorphism and W is any subset of a free group, then $\phi W(G_1) \subset W(G_2)$ since $\phi\omega \in \Omega(G_2)$ for any $\omega \in \Omega(G_1)$. Consequently, there is induced a unique homomorphism ϕ_* such that

$$\begin{array}{ccc} G_1 & \xrightarrow{\phi} & G_2 \\ \downarrow & & \downarrow \\ G_1/W(G_1) & \xrightarrow{\phi_*} & G_2/W(G_2) \end{array}$$

is a consistent diagram. It is straightforward to prove that

(4.1) (a) *If ϕ is the identity, so is ϕ_*.* (b) *Given the composition $G_1 \xrightarrow{\phi} G_2 \xrightarrow{\psi} G_3$, then $(\psi\phi)_* = \psi_*\phi_*$.* (c) *If ϕ is onto, so is ϕ_*.* (d) *If ϕ is an isomorphism onto, so is ϕ_*.*

For example, if (a) and (b) have been verified, (d) follows by simply observing

$$\begin{aligned} \text{identity} &= (\phi\phi^{-1})_* = \phi_*(\phi^{-1})_* \\ &= (\phi^{-1}\phi)_* = (\phi^{-1})_*\phi_*. \end{aligned}$$

Notice that if (d) is altered by the omission of the word "onto," the result is false.

Using (4.1), we can prove a result that was promised in Chapter III.

(4.2) *If* $\mathfrak{m}$ *and* $\mathfrak{n}$ *are distinct cardinal numbers, then the free groups of rank* $\mathfrak{m}$ *and* $\mathfrak{n}$ *are not isomorphic.*

Proof. Consider free groups $F_\mathfrak{m}$ and $F_\mathfrak{n}$ of rank $\mathfrak{m}$ and $\mathfrak{n}$ respectively, and assume that they are isomorphic. Then by (4.1) we must have an isomorphism of $F_\mathfrak{m}/F_\mathfrak{m}^2$ onto $F_\mathfrak{n}/F_\mathfrak{n}^2$. However the elements of these groups can be exhibited explicitly and counted; since $F_\mathfrak{m}/F_\mathfrak{m}^2$ is abelian its elements are just the products $x_1^{\delta_1}x_2^{\delta_2}\cdots x_\mathfrak{m}^{\delta_\mathfrak{m}}$, where $x_1, \cdots, x_\mathfrak{m}$ is a basis for $F_\mathfrak{m}$, $\delta_i = 0$ or 1, and only a finite number of the exponents δ_i are different from 0. Thus the number of elements of $F_\mathfrak{m}/F_\mathfrak{m}^2$ is just the number of finite subsets of a set of cardinality $\mathfrak{m}$; it is $2^\mathfrak{m}$ for finite $\mathfrak{m}$ and $\mathfrak{m}$ for infinite $\mathfrak{m}$. If $F_\mathfrak{m}/F_\mathfrak{m}^2 \approx F_\mathfrak{n}/F_\mathfrak{n}^2$, then $\mathfrak{m}$ and $\mathfrak{n}$ must be both finite or both infinite, and hence $\mathfrak{m} = \mathfrak{n}$.

From (4.1) there follows the most elementary of all necessary conditions in the isomorphism problem:

(4.3) *In order that* G_1 *and* G_2 *be isomorphic it is necessary that their commutator quotient groups* $G_1/[G_1,G_1]$ *and* $G_2/[G_2,G_2]$ *be isomorphic.*

The commutator quotient group $G/[G,G]$ of any group G is the largest abelian group which is a homomorphic image of G. This idea is expressed rigorously in the following way. Consider an arbitrary homomorphism θ of G into an abelian group K. Then, there exists a unique homomorphism θ' mapping $G/[G,G]$ into K which is consistent with θ and the abelianizer $\mathfrak{a}\colon G \to G/[G,G]$.

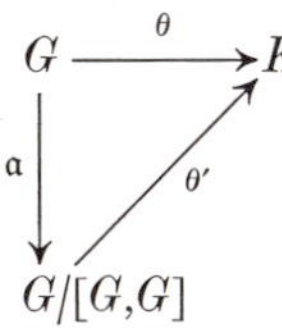

$$\begin{array}{ccc} G & \xrightarrow{\theta} & K \\ {\scriptstyle\mathfrak{a}}\downarrow & \nearrow{\scriptstyle\theta'} & \\ G/[G,G] & & \end{array} \qquad \theta = \theta'\mathfrak{a}$$

To prove this assertion, consider an arbitrary commutator

$$[g_1,g_2] = g_1g_2g_1^{-1}g_2^{-1}, \quad g_1,g_2 \in G.$$

Since K is abelian,

$$\theta[g_1,g_2] = [\theta g_1,\theta g_2] = 1,$$

and, therefore the consequence of the commutators of G is contained in the kernel of θ. The group $[G,G]$ generated by the commutators of G is ipso facto contained in their consequence; hence θ' is well-defined by

$$\theta'\mathfrak{a}g = \theta g, \qquad g \in G.$$

The uniqueness of θ' follows trivially. Notice that since $[G,G]$ is a normal subgroup of G it actually equals the consequence of the commutators of G. The result just proved is succinctly summarized in the statement

(4.4) *Any homomorphism of a group into an abelian group can be factored through the commutator quotient group.*

(4.5) *If a group G is generated by $g_1, g_2, \cdots$, then its commutator subgroup $[G,G]$ is the consequence of the commutators $[g_i,g_j]$, $i, j = 1, 2, \cdots$.*

Proof. The consequence K of the commutators $[g_i,g_j]$ is contained in every normal subgroup of G that contains $\{[g_i,g_j]\}$. Hence $K \subset [G,G]$. To prove the converse we have to show that the commutator $[u,v]$ of any two elements of G lies in K.

For every element $g \in G$, let $l(g)$ denote the smallest non-negative integer n for which there exist $\epsilon_1, \epsilon_2, \cdots, \epsilon_n = \pm 1$ such that $g = \Pi_{k=1}^{n} g_k^{\epsilon_k}$. Obviously $l(g) = 0$ if and only if $g = 1$. Our proof of the above proposition is by induction on $l(u) + l(v)$, If either $l(u) = 0$ or $l(v) = 0$ then $[u,v] = 1 \in K$. If $l(u) = l(v) = 1$ the commutator $[u,v]$ is one of the following:

$$\begin{aligned} &[g_i,g_j], \\ &[g_i^{-1},g_j] = g_i^{-1}[g_j,g_i]g_i, \\ &[g_i,g_j^{-1}] = g_j^{-1}[g_j,g_i]g_j, \\ &[g_i^{-1},g_j^{-1}] = g_i^{-1}g_j^{-1}[g_i,g_j]g_jg_i, \end{aligned}$$

and each of these must belong to K. Assume next that either $l(u)$ or $l(v)$ is greater than 1. As a result of the identity

$$[u,v] = [v,u]^{-1},$$

we may assume that it is $l(u)$ that is greater than 1. Then $u = u_1u_2$ where $l(u_1) < l(u)$ and $l(u_2) < l(u)$. By the inductive hypothesis

$$[u_1u_2,v] = u_1[u_2,v]u_1^{-1}[u_1,v] \in K,$$

and this completes the proof.

(4.6) *If $(\mathbf{x} : \mathbf{r})$ is any group presentation, then $(\mathbf{x} : \mathbf{r} \cup \{[x_i,x_j], i,j = 1, 2, \cdots\})$ is a presentation of the abelianized group of $|\mathbf{x} : \mathbf{r}|$.*

Proof. Let γ denote the canonical homomorphism of the free group $F(\mathbf{x})$ onto the factor group $|\mathbf{x} : \mathbf{r}|$. The abelianizer of $|\mathbf{x} : \mathbf{r}|$ is denoted as before by $\mathfrak{a}$. We now have to show that the kernel of $\mathfrak{a}\gamma$ is the consequence K of $\mathbf{r} \cup \{[x_i,x_j]\}$. That K is contained in this kernel is trivial. To prove the reverse inclusion select any element u of $F(\mathbf{x})$ such that $\alpha\gamma u = 1$. By (4.5) γu is contained in the consequence of $\{[\gamma x_i, \gamma x_j]\} = \{\gamma[x_i, x_j]\}$. But, by (1.1), this is the image under γ of the consequence of $\{[x_i, x_j]\}$. Thus $\gamma u = \gamma v$ for some consequence v of $\{[x_i, x_j]\}$. Hence $u = vw$ for some consequence w of $\mathbf{r}$, and the proof is complete.

5. Free abelian groups. In the theory of abelian groups, one encounters another kind of free group, which is quite analogous to, but different from, the free group as it is defined in this book. Specifically, a *free abelian group of rank n* is any group which is isomorphic to the abelianized group of a free group of rank n. Since the only commutative free groups are those of rank 0 and 1, i.e., the trivial groups and the infinite cyclic groups, it is clear that the two notions overlap but do not coincide. Generally speaking, a free abelian group is not a free group. For both types, however, the rank is a complete invariant. Thus,

(5.1) *Two free abelian groups are isomorphic if and only if they have the same rank.*

It is not hard to construct a proof based on the same result for free groups, (4.3), and the technique used in proving (4.2) (cf. Exercise 2 below).

There is an abstract characterization of the free abelian groups which is entirely analogous to that of the free groups. A generating set E of elements of an abelian group G is a *basis* if, given any abelian group H, any function $\phi: E \to H$ can be extended to a homomorphism of G into H. Then,

(5.2) *An abelian group is free abelian if and only if it has a basis.*

A proof based on (4.4) is straightforward.

Because of (4.6) it is an easy matter to give simple presentations of the free abelian groups. For example $(x,y : xy = yx)$ and $(x,y,z : xy = yx, yz = zy, zx = xz)$ present the free abelian groups of rank 2 and 3 respectively.

EXERCISES

1. Prove the following addendum to (4.1): If ϕ is an isomorphism into, then ϕ_* need *not* be an isomorphism into. (One solution: $G_1 = (x:)$, $G_2 = (u,v:)$, $\phi(x) = uvu^{-1}v^{-1}$, $W(G) = [G,G]$.)

2. If F_n denotes the free group and A_n the free abelian group of rank n, show that $F_n/F_n{}^2 \approx A_n/A_n{}^2$. Deduce (5.1) from this.

3. Prove (5.2).

4. How many different homomorphisms are there of the free group of rank 2 onto the cyclic group of order 4?

5. Show that the presentations $(a,b : a^2 = 1, b^3 = 1, ab = ba)$ and $(c : c^6 = 1)$ describe the same group.

6. Show that the group presented by $(x,y : xy^2 = y^3x, yx^2 = x^3y)$ is the trivial group. (This is a hard problem.)

7. In the presentation $(a,b,c,d : b = c^{-1}ac, c = dbd^{-1}, d = a^{-1}ca, a = bdb^{-1})$ verify that any one of the relations is a consequence of the others.

8. Show that the presentation $(a,b : a^3 = 1, b^2 = 1, ab = ba^2)$ describes the symmetric group of degree 3.

9. Describe the word subgroup $W(G)$ of an arbitrary group G for (i) $W = xy$, (ii) $W = x^6y^9$, (iii) $W = xyxy^{-1}$.

10. Is theorem (1.1) necessarily true if ϕ is not onto?

11. Show that (4.5) is false if the words "consequence of" are replaced by the words "subgroup generated by".

CHAPTER V

Calculation of Fundamental Groups

Introduction. It was remarked in Chapter II that a rigorous calculation of the fundamental group of a space X is rarely just a straightforward application of the definition of $\pi(X)$. At this point the collection of topological spaces whose fundamental groups the reader can be expected to know (as a result of the theory so far developed in this book) consists of spaces topologically equivalent to the circle or to a convex set. This is not a very wide range, and the purpose of this chapter is to do something about increasing it. The techniques we shall consider are aimed in two directions. The first is concerned with what we may call spaces of the same shape. Figures 14, 15, and 16 are examples of the sort of thing we have in mind. From an understanding of the fundamental group as formed from the set of classes of equivalent loops based at a point, it is geometrically apparent that the spaces shown in Figure 14 below have the same, or isomorphic, fundamental groups.

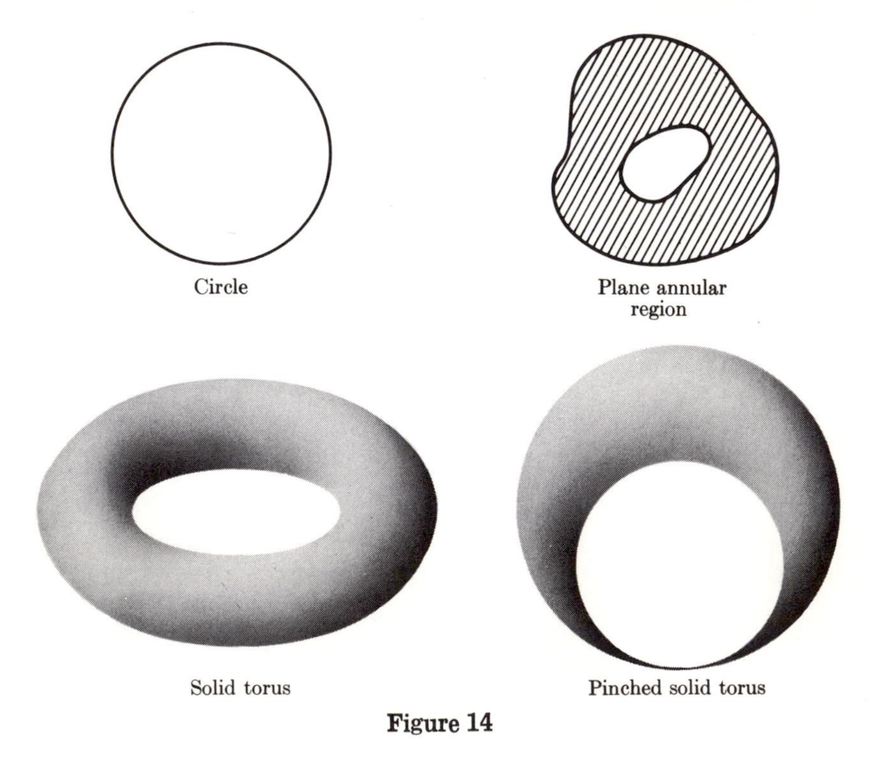

Figure 14

Similarly for Figure 15.

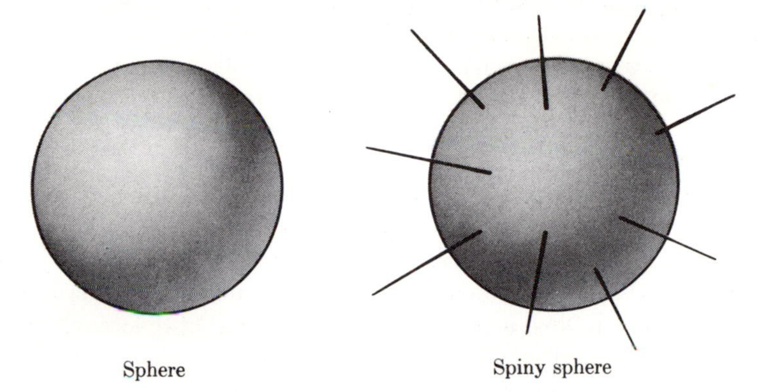

Figure 15

And again in Figure 16.

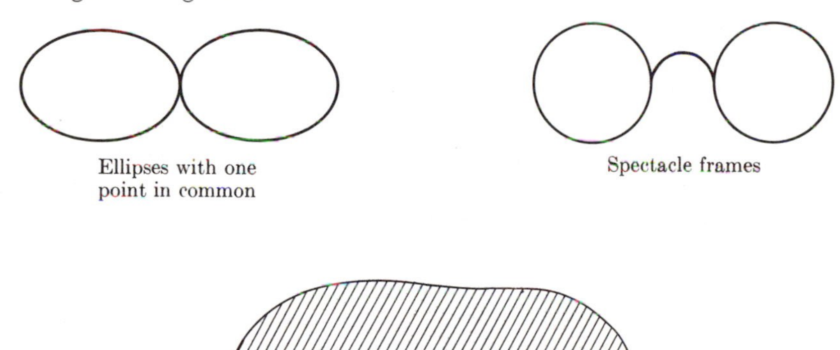

Double plane annulus

Figure 16

We have purposely chosen these examples so that no two are topologically equivalent. (This fact is not obvious.[1]) Nevertheless, all spaces included in the

[1] Most of the above spaces under the same figure are distinguishable from one another because of the fact that the dimension of a topological space in the neighborhood of a point is a topological invariant, i.e., under a homeomorphism the local dimension for any point is the same as that of its image under the homeomorphism. See W. Hurewicz and H. Wallman, *Dimension Theory*, (Princeton University Press, Princeton, New Jersey, 1948).

same figure have isomorphic fundamental groups and, in some sense, are of the same shape. In the first two sections of this chapter we shall study some of the mathematical terminology used in describing this concept precisely and its relation to the fundamental group. The definitions introduced will be those of *retraction*, *deformation*, *deformation retract*, and *homotopy type*. The relation of homotopy type is the rigorous replacement of our vague notion of spaces of the same shape.

The second over-all topic of this chapter is the calculation of the fundamental groups of spaces which are built up in a systematic way from simpler spaces whose fundamental groups are known. As a simple example, consider the space consisting of the union of two circles X and Y which have a single point p in common. Now it is very reasonable—and also correct—to guess that $\pi(X \cup Y)$ is the free group on two generators: one generating $\pi(X)$ and the other $\pi(Y)$. However, this conclusion is certainly not an obvious corollary of any techniques so far developed. Clearly, it would be of tremendous importance to have a general procedure for cementing together the fundamental groups which correspond to the spaces that are being joined. For a wide variety of spaces such a procedure exists; it is derived from the *van Kampen theorem*. Most spaces encountered in topology, specifically the so-called *complexes*, do exhibit a decomposition as the union of structurally simple subsets. By repeated application of the van Kampen theorem to these components the collection of spaces whose fundamental groups are readily calculable is enormously enlarged. In Section 3 we shall give a precise statement of this all important tool and discuss its application in several examples. A proof for it appears in Appendix III.

1. Retractions and deformations. A *retraction* of a topological space X onto a subspace Y is a continuous mapping $\rho\colon X \to Y$ such that, for any p in Y, $\rho(p) = p$. A space Y is called a *retract* of X if there exists a retraction $\rho\colon X \to Y$.

As an example, consider the square Q in 2-dimensional Euclidean space R^2 defined by $0 \leq x \leq 1$, $0 \leq y \leq 1$. A retraction of Q onto the edge E defined by $0 \leq x \leq 1$, $y = 0$ is given by

$$\rho(x,y) = (x,0), \qquad 0 \leq x,y \leq 1.$$

By restricting the domain of this function to the set $\dot{Q}$ consisting of all (x,y) in Q such that $xy(x - 1)(y - 1) = 0$ (at least one of the factors must equal zero), we obtain a retraction of the boundary of the square onto the edge E. In addition, the origin $(0,0)$ is a retract of the square, of its boundary $\dot{Q}$, and also of the edge E. The retraction, with domain suitably chosen in each case, is given by the function

$$\rho(x,y) = (0,0).$$

More generally, for any point p of an arbitrary topological space X, the constant mapping $\rho\colon X \to p$ is a retraction. A point is therefore a retract of any space that contains it.

It follows from the above that any interior point p of a closed circular disc D is a retract of the disc. By projecting radially outward from p onto the boundary $\dot{D}$ of the disc, we see that $\dot{D}$ is a retract of the complementary space $D - p$. Analogous results hold in 3 dimensions. If p is the center of D and if D is rotated about a diameter to form a 3-cell B, then p is a retract of B, and the boundary sphere is a retract of $B - p$. If the disc D is rotated about a line lying in its exterior, D describes a solid torus V and p describes a circle C. We may conclude that C, forming the core of the solid torus, is a retract of V. Similarly the torus that constitutes the boundary of V is a retract of $V - C$.

It is equally important, of course, to give examples of subspaces which are *not* retracts of their containing spaces. These are also readily available. The boundary of a square is not a retract of the square. Similarly, $\dot{D}$ is not a retract of the disc D. There is no retraction of 3-space R^3 onto the solid torus V nor onto its core C. The equator is not a retract of the surface of the earth. Well, how do we know? How can one possibly prove the nonexistence of a retraction? An answer is given in the following theorem.

(1.1) *If $\rho\colon X \to Y$ is a retraction and X is pathwise connected, then, for any basepoint $p \in X$, the induced homomorphism $\rho_*\colon \pi(X,p) \to \pi(Y,\rho p)$ is onto.*

Proof. It was observed in Chapter II (cf. (4.6) and accompanying discussion) that the algebraic properties of any homomorphism induced by a continuous mapping of a pathwise connected space are independent of the choice of basepoint. For this reason, it is sufficient to check (1.1) for a basepoint $p \in Y$. Consider the induced homomorphisms

$$\pi(Y,p) \xrightarrow{i_*} \pi(X,p) \xrightarrow{\rho_*} \pi(Y,p),$$

where $i\colon Y \to X$ is the inclusion. Since ρ is a retract, the composition ρi is the identity. Hence (cf. (4.5), Chapter II), we have $(\rho i)_* = \rho_* i_* =$ identity, and it follows that ρ_* is onto, and we are finished.

The continuous image of a pathwise connected space is pathwise connected. Thus, the space Y appearing in (1.1) is connected too. Without the provision that X be pathwise connected, (1.1) would be false.

The square, the disc, and the space R^3 are convex sets and hence possess trivial fundamental groups. It is true that a proof of the fact that the fundamental group of the sphere is trivial is not given until the end of this chapter. However, the fact that all loops on the sphere having a common basepoint can be contracted to that point certainly sounds plausible. Incidentally, the word "sphere" alone always means the surface of the solid cell. Thus, in all the examples mentioned in the paragraph preceding (1.1), the fundamental group of the containing space X is the trivial group. The subspace Y, however, in each example has an infinite cyclic fundamental group. A group containing only one element can obviously not be mapped onto one containing more than one, and the contentions of that paragraph are therefore proved.

We turn now to the notion of a deformation. The intuitive idea here is almost self-explanatory. A topological space X is deformable into a subspace Y if X can be continuously shrunk into Y. The words "into" and "onto" have their usual meaning. If the result of the shrinking is a set not only contained in Y but equal to Y, then we say X is deformable onto Y. For example, a square is deformable onto one of its edges. The appropriate definitions are: A *deformation* of a topological space X is a family of mappings h_s: $X \to X$, $0 \leq s \leq 1$, such that h_0 is the identity, i.e., $h_0(p) = p$ for all p in X, and the function h defined by $h(s,p) = h_s(p)$ is simultaneously continuous in the two variables s and p.[2] A *deformation of a space X into* (or *onto*) *a subspace Y* is a deformation $\{h_s\}$ of X such that the image h_1X is contained in (or equal to) Y. We say that X is *deformable* into, or onto, a subspace Y if such a deformation exists.

A square, for example, can be deformed onto an edge; a deformation in R^2 is given by

$$\text{(i)} \quad h_s(x,y) = (x, (1-s)y), \qquad 0 \leq x,y,s \leq 1.$$

The family of functions

$$\text{(ii)} \quad h_s(x,y) = ((1-s)x, (1-s)y), \qquad 0 \leq x,y,s \leq 1,$$

is a deformation of the square onto the corner (0,0). The disc D defined with respect to polar coordinates by the inequality $0 \leq r \leq 1$ is deformed onto its center by the functions

$$\text{(iii)} \quad h_s(r,\theta) = (r(1-s), \theta), \qquad \begin{cases} 0 \leq r,s \leq 1, \\ 0 \leq \theta < 2\pi. \end{cases}$$

The complementary space consisting of all points of D except the center can be deformed onto the boundary $\dot{D}$ of the disc. The deformation is given by

$$\text{(iv)} \quad h_s(r,\theta) = (r(1-s) + s, \theta), \qquad \begin{cases} 0 < r \leq 1, \\ 0 \leq s \leq 1, \\ 0 \leq \theta < 2\pi. \end{cases}$$

The reader should check that for $0 < r \leq 1$, $0 \leq s \leq 1$, the inequalities $r \leq r(1-s) + s \leq 1$ are satisfied. Notice also that (iv) cannot be extended to a deformation of D onto $\dot{D}$. As before, we may extend our considerations of a disc into 3 dimensions and conclude, for example, that a solid torus V

---

[2] A generalization is the definition of a deformation of X *in a containing space* Z as a family of mappings h_s: $X \to Z$, $0 \leq s \leq 1$, satisfying $h_0(p) = p$ for all p in X and the condition of simultaneous continuity. The more restricted definition of deformation given above is suitable for our purposes.

can be deformed onto a circle C forming its core, and that $V - C$ can be deformed onto the torus that forms the boundary of V.

A deformation of a disc onto its boundary can be constructed as follows: Let the disc D be defined with respect to polar coordinates to be the set of all points (r,θ) that satisfy $0 \leq r \leq \sin\theta$. (See Figure 17.) The boundary $\dot{D}$ is

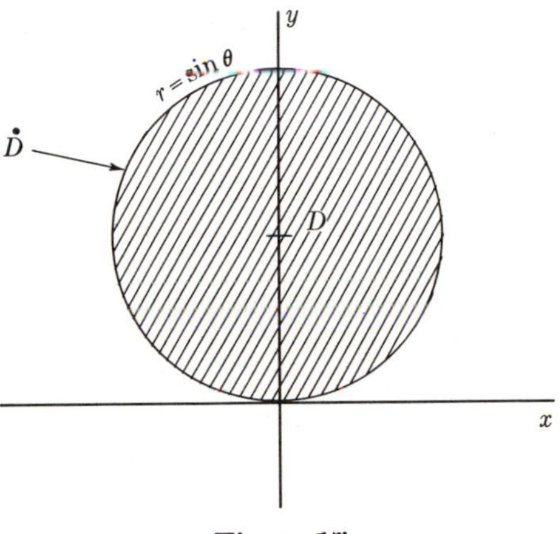

Figure 17

the set of points defined by $r = \sin\theta$. The deformation occurs in two steps: In the interval $0 \leq s \leq \frac{1}{2}$, the entire disc is shrunk to the origin p. Then, in $\frac{1}{2} \leq s \leq 1$, points are moved along the boundary $\dot{D}$ in the counterclockwise direction with increasing r and s. Specifically, for any ordered pair (r,θ) such that $0 \leq r \leq \sin\theta$, we set

$$h_s(r,\theta) = \begin{cases} (r(1 - 2s), \theta), & \text{for } 0 \leq s \leq \frac{1}{2}, \\ (\sin \pi r(2s - 1), \pi r(2s - 1)), & \text{for } \frac{1}{2} \leq s \leq 1. \end{cases}$$

It is not hard to convince oneself that this family of functions satisfies the requirement of simultaneous continuity prescribed by the definition of a deformation. This example is particularly interesting because we know from (1.1) that *the boundary $\dot{D}$ is not a retract of the disc D.*

We have just seen an example of a topological space X which can be deformed onto a subspace which is not a retract of X. It is natural to ask whether, conversely, there exist retracts which cannot be obtained by deformation. The answer is yes, and a good tool for finding examples is Theorem (1.3) below. We prove first, as a lemma,

(1.2) *If $\{h_s\}$, $0 \leq s \leq 1$, is a deformation of X, then, for any basepoint p in X, the homomorphism $(h_1)_*: \pi(X,p) \to \pi(X,h_1(p))$ is an isomorphism onto.*

Proof. We define a path a with initial point p and terminal point $q = h_1(p)$ by the formula

$$a(t) = h_t(p), \qquad 0 \leq t \leq 1.$$

Let the equivalence class of paths in X containing a be denoted by α. We shall show that, for any β in $\pi(X,p)$,

$$(h_1)_*(\beta) = \alpha^{-1} \cdot \beta \cdot \alpha,$$

and the result then follows from (3.1), Chapter II. Consider therefore an arbitrary element β in $\pi(X,p)$ and representative loop b in β. Set

$$k_s(t) = h_s(b(t)), \qquad 0 \leq s \leq 1,\ 0 \leq t \leq \| b \|.$$

The collection $\{k_s\}$ is certainly a continuous family of loops; its domain is conveniently pictured in Figure 18.

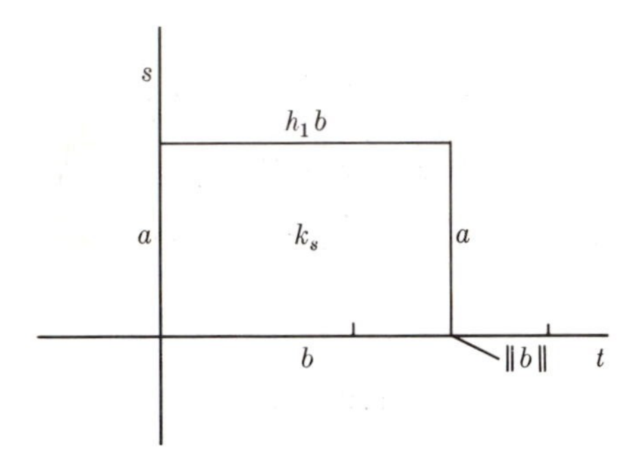

Figure 18

Another continuous family of paths is defined as follows and represented in Figure 19.

$$j_s(t) = \begin{cases} a(1 - (t - s)), & 0 \leq s \leq t \leq 1, \\ q, & 0 \leq t \leq s \leq 1. \end{cases}$$

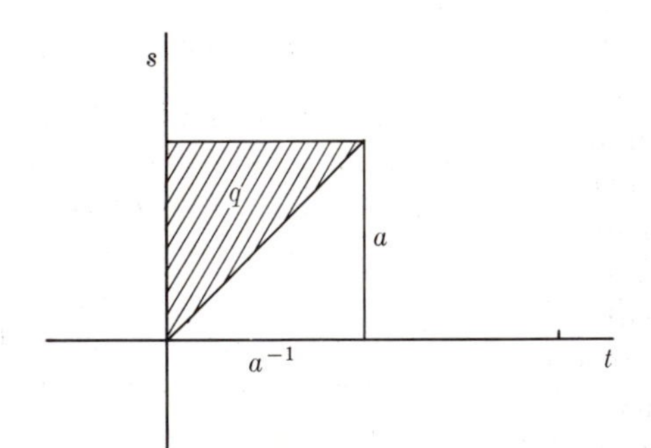

Figure 19

A final continuous family of paths is represented in Figure 20 and defined by

$$l_s(t) = \begin{cases} a(s + t), & s + t \leq 1,\ 0 \leq s,t \leq 1, \\ q, & s + t \geq 1,\ 0 \leq s,t \leq 1. \end{cases}$$

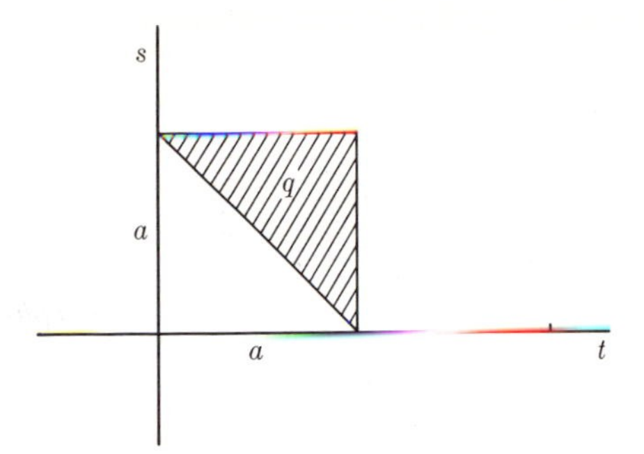

Figure 20

The product family $\{j_s \cdot k_s \cdot l_s\}$ is clearly defined and is a continuous family of paths. It is represented in Figure 21.

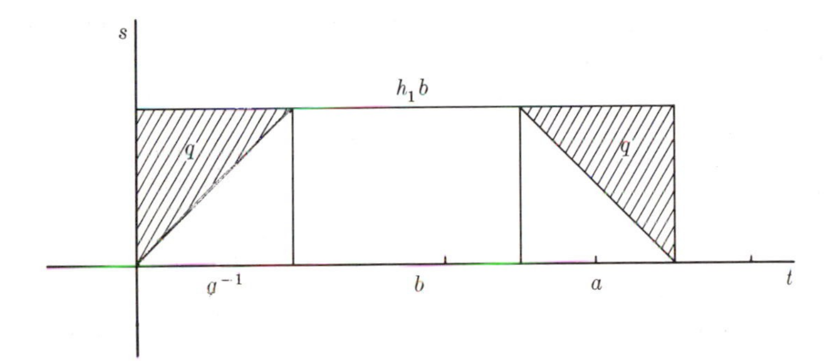

Figure 21

Since $\{j_s \cdot k_s \cdot l_s\}$ is a fixed-endpoint family, we have the equivalences

$$a^{-1} \cdot b \cdot a \simeq c \cdot h_1 b \cdot c,$$

where c is the constant path at q. The equivalence class containing any constant path is an identity (cf. (2.6), Chapter II). Hence,

$$\begin{aligned} \alpha^{-1} \cdot \beta \cdot \alpha &= [a^{-1} \cdot b \cdot a] = [h_1 b] \\ &= (h_1)_* \beta, \end{aligned}$$

and the proof is complete.

An arbitrary continuous mapping $f\colon X \to Y$ of a topological space X into a subspace Y is said to be *realizable by a deformation of* X if there exists a deformation $\{h_s\}$, $0 \leq s \leq 1$, of X such that $h_1 = if$, where $i\colon Y \to X$ is the inclusion mapping. As a corollary of (1.2), we have the following theorem.

(1.3) *If a continuous mapping $f: X \to Y$ is realizable by a deformation of X, then, for any basepoint p in X, the induced homomorphism $f_*: \pi(X,p) \to \pi(Y,fp)$ is an isomorphism into.*

Proof. Let $i: Y \to X$ be the inclusion mapping and $\{h_s\}$ a deformation of X such that $h_1 = if$. Since $(h_1)_* = i_* f_*$ and since $(h_1)_*$ is an isomorphism by (1.2), we may conclude that f_* is also an isomorphism. Note that f_* is, in general, not an onto mapping even though $(h_1)_*$ is.

Obviously, a space X is deformable into a subspace Y if and only if there exists a mapping $f: X \to Y$ which is realizable by a deformation of X. Consequently, by virtue of Theorem (1.3), it is easy to find examples of retracts which cannot be obtained by deformation. As we have seen, an edge E of a square Q is a retract of the boundary $\dot{Q}$ of the square. However, $\dot{Q}$ cannot be deformed into E. The fundamental group $\pi(\dot{Q})$ is infinite cyclic and $\pi(E)$ is trivial, so no mapping of $\pi(\dot{Q})$ into $\pi(E)$ can be an isomorphism. Similarly, $\dot{Q}$ cannot be deformed onto a point. In contrast to a similar statement we observed to hold for retracts, it is certainly false that an arbitrary space X is deformable onto any point in X.

We are now ready to combine the notions of retraction and deformation into a single definition. A subspace Y of a topological space X is a *deformation retract* of X if there exists a retraction $\rho: X \to Y$ which is realizable by a deformation of X.

Since h_1 defines a retraction in each of the formulas (i), (ii), (iii), (iv), each exhibits a deformation retract. Thus, both an edge of a square and a corner point are deformation retracts of the square. An interior point p of a disc D is a deformation retract of D, and the boundary of D is a deformation retract of $D - p$. In the following theorem, which is a direct corollary of (1.1) and (1.3), we obtain an important property of deformation retracts.

(1.4) *If a subspace Y is a deformation retract of a pathwise connected topological space X, then $\pi(X)$ is isomorphic to $\pi(Y)$.*

Notice in this theorem that Y must also be pathwise connected. A more informative statement of (1.4) is the following:

If X is pathwise connected and the retraction $\rho: X \to Y$ is realizable by a deformation and if $i: Y \to X$ is the inclusion mapping, then, for any points p in X and q in Y, both induced homomorphisms

$$\rho_*: \pi(X,p) \to \pi(Y,\rho p),$$
$$i_*: \pi(Y,q) \to \pi(X,q),$$

are isomorphisms onto.

The first result is a direct corollary of (1.1) and (1.3). To prove the second, consider the mappings

$$\pi(X,q) \xrightarrow{\rho_*} \pi(Y,q) \xrightarrow{i_*} \pi(X,q).$$

We have already noted that ρ_* is an isomorphism onto. Moreover, we know that there exists a deformation $\{h_s\}$ of X such that $h_1 = i\rho$. By (1.2), $(h_1)_* = i_*\rho_*$ is an isomorphism onto, and it follows immediately that i_* is also an isomorphism onto. The big difference between the two statements of Theorem (1.4) is that the latter not only says that two groups are isomorphic, but also says explicitly what the two relevant isomorphisms are.

The concepts of retraction, deformation, and deformation retract can be nicely summarized with the example of a square Q. Since there exists a homeomorphism of Q onto the circular disc D that carries the boundary $\dot{Q}$ onto the boundary $\dot{D}$, we may conclude that the square, like the disc, may be deformed onto its boundary. We thus have the following diagram:

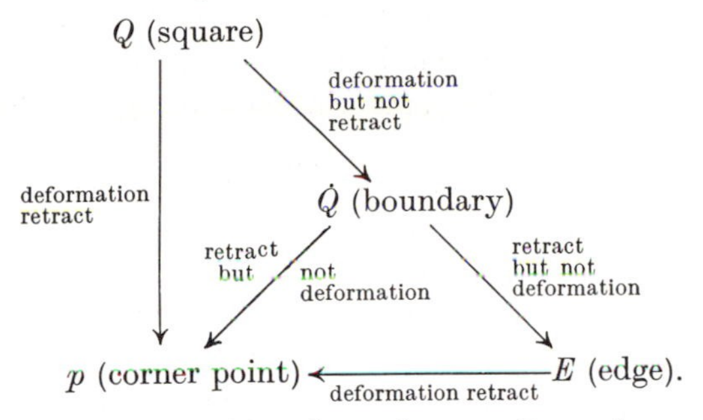

Let the closed disc defined by the polar coordinate inequality $0 \leq r \leq 1$ be denoted by D, and its center by p. The open disc, defined by $0 \leq r < 1$, is denoted by $\mathring{D}$. Notice that $\mathring{D} = D - \dot{D}$. The deformation (iv) of $D - p$ onto $\dot{D}$ can be extended as follows: For any r,s,θ such that $0 < r$, $0 \leq s \leq 1$, $0 \leq \theta < 2\pi$, set

$$h_s(r,\theta) = \begin{cases} (r(1-s)+s,\ \theta), & \text{if } \ 0 < r \leq 1, \\ (r,\theta), & \text{if } \ 1 \leq r. \end{cases}$$

This extended deformation shows that the complement $R^2 - \mathring{D}$ of the open disc in the plane is a deformation retract of the punctured plane $R^2 - p$. By rotating the disc D about an axis lying outside D, as we have done before, we obtain a solid torus V, whose surface and interior are denoted by $\dot{V}$ and $\mathring{V}$ respectively. The point p describes a circle C under the rotation, and it is obvious that the torus $\dot{V}$ is a deformation retract of the complement $V - C$. Consider next a topological imbedding of the closed solid torus V into the 3-dimensional space R^3. The image of C under the imbedding is a knot K. The knotted torus which is the image of V and which contains K as a core, we denote by W. Its surface and interior are denoted by $\dot{W}$ and $\mathring{W}$, respectively. It can be proved that the imbedding of V into R^3 must carry $\dot{V}$ onto $\dot{W}$ and $\mathring{V}$ onto $\mathring{W}$. Consequently, it follows that $\dot{W}$ is a deformation retract of $W - K$; since the points of $\dot{W}$ remain fixed throughout the deformation, we may extend the mapping to all of $R^3 - K$ and conclude that $R^3 - \mathring{W}$ is a deformation retract of $R^3 - K$. Thus, $R^3 - K$ and $R^3 - \mathring{W}$ have isomorphic fundamental groups.[3] It is obvious that any knot of tame type is the core of such an open toroidal neighborhood. It has been proved[4] that conversely, any knot which possesses such a toroidal neighborhood is tame.

The following theorem, with which we conclude this section, is perhaps an unexpected result. At first glance, one is very likely to guess that it is false. As a matter of fact, the proof is almost a triviality.

(1.5) *If the space X can be deformed into Y and if there also exists a retraction ρ of X onto Y, then Y is a deformation retract of X; moreover, ρ can be realized by a deformation.*

Proof. Let $\{h_s\}$ be a deformation of X into Y. We define a new deformation $\{k_s\}$ of X as follows. For any point p in X,

$$k_s(p) = \begin{cases} h_{2s}(p), & 0 \leq s \leq \frac{1}{2}, \\ \rho h_{2-2s}(p), & \frac{1}{2} \leq s \leq 1. \end{cases}$$

Then, $k_0(p) = h_0(p) = p$. The condition of simultaneous continuity in s and p is satisfied because the two definitions of $k_{\frac{1}{2}}$ agree. Using the top line, we get $k_{\frac{1}{2}}(p) = h_1(p)$. Since $h_1(p)$ is by assumption in Y, we obtain from the second line $k_{\frac{1}{2}}(p) = \rho h_1(p) = h_1(p)$. The retraction ρ is realized by the deformation because $k_1(p) = \rho h_0(p) = \rho(p)$.

2. Homotopy type. Topological spaces X and Y are of the same *homotopy type* if there exists a finite sequence

$$X = X_0, X_1, \cdots, X_n = Y$$

of topological spaces such that, for each $i = 1, \cdots, n$, either X_i is topologically equivalent to X_{i-1}, or X_i is a deformation retract of X_{i-1}, or vice-versa.[5] The relation of belonging to the same homotopy type is obviously an equivalence relation. From (1.4) above, and (4.7) of Chapter II, we conclude that

(2.1) *If X and Y are pathwise connected spaces of the same homotopy type, then $\pi(X)$ is isomorphic to $\pi(Y)$.*

Any point in a convex set C is a deformation retract of C. It follows that

---

[3] However, the fundamental groups of $R^3 - K$ and $R^3 - W$ may not be isomorphic (the torus W may be "horned"); See J. W. Alexander, "An Example of a Simply Connected Surface Bounding a Region which is not Simply Connected," *Proceedings of the National Academy of Sciences*, Vol. 10 (1924), pp. 8–10.

[4] E. Moise, "Affine Structures in 3-Manifolds, V. The Triangulation Theorem and Hauptvermutung," *Ann. of Math.* Vol. 56 (1952), pp. 96–114.

[5] For the usual definition of homotopy type see P. J. Hilton, *An Introduction to Homotopy Theory*, Cambridge Tracts in Mathematics and Mathematical Physics, No. 43 (Cambridge University Press, Cambridge, 1953). For proof that Hilton's definition is the same as ours see R. H. Fox, "On Homotopy Type and Deformation Retracts," *Ann. of Math.* Vol. 44 (1943), pp. 40–50.

any convex set is of the same homotopy type as a point. Thus our collection of spaces of known fundamental group may be characterized as all those of the homotopy type of a circle or a point. Homotopy type is one of the most important equivalence relations in algebraic topology; most of the algebraic invariants are invariants of homotopy type. It is a much weaker relation than that of topological equivalence. Each of the spaces pictured in Figures 14, 15, and 16 is of the same homotopy type as the others in the same figure.

3. The van Kampen theorem. The formulation of this important result which appears in (3.1) does not, at first glance, look like a computational aid to finding fundamental groups. There are, however, distinct advantages to this abstract approach. One is conceptual simplicity: The statement of (3.1) reflects only the essential algebraic structure of the theorem and, for this reason, is the easiest and clearest one to prove. In addition, the important corollaries ((3.2), (3.3), and (3.4)) needed in the next chapter follow most easily and directly from the abstract presentation. The classical formulation of the van Kampen theorem in terms of generators and relators is derived and given in (3.6).

Let X be a topological space which is the union $X = X_1 \cup X_2$ of open subsets X_1 and X_2 such that X_1, X_2, and $X_0 = X_1 \cap X_2$ are all pathwise connected and nonvoid. Since the intersection X_0 is nonvoid, it follows that the space X is pathwise connected. We select a basepoint $p \in X_0$ and set $G = \pi(X,p)$ and $G_i = \pi(X_i,p)$, $i = 0, 1, 2$. The homomorphisms induced by inclusion mappings form the consistent diagram

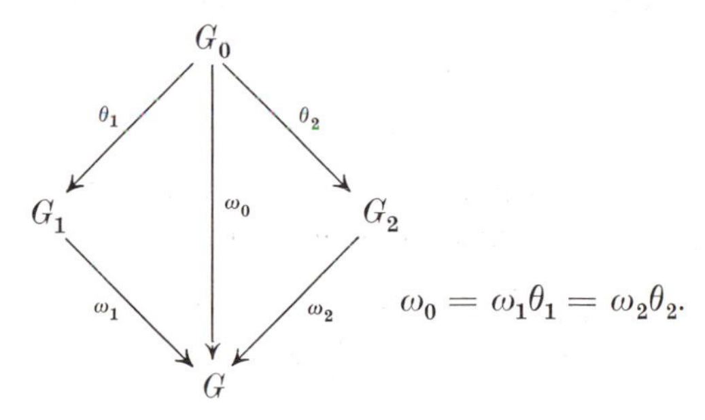

(3.1) THE VAN KAMPEN THEOREM. *The image groups $\omega_i G_i$, $i = 0, 1, 2$, generate G. Furthermore, if H is an arbitrary group and $\psi_i\colon G_i \to H, i = 0, 1, 2$, are homomorphisms which satisfy $\psi_0 = \psi_1\theta_1 = \psi_2\theta_2$, then there exists a unique homomorphism $\lambda\colon G \to H$ such that $\psi_i = \lambda\omega_i$, $i = 0, 1, 2$.*

A proof is given in Appendix III. Notice that, in view of the consistency relation $\omega_0 = \omega_1\theta_1 = \omega_2\theta_2$, the assertion that $\omega_1 G_1$ and $\omega_2 G_2$ generate G is fully equivalent to and may replace the first sentence in (3.1). An immediate corollary is then

(3.2) *If X_1 and X_2 are simply-connected, then so is $X = X_1 \cup X_2$.*

As will be shown in detail among other examples at the end of this chapter, the next corollary of the van Kampen theorem solves the problem of determining the fundamental group of two circles joined at a point (cf. Figure 22).

(3.3) *If G_0 is trivial and G_1 and G_2 are free groups with free bases* $\{\alpha_1, \alpha_2, \cdots\}$ *and* $\{\beta_1, \beta_2, \cdots\}$, *respectively, then G is free and* $\{\omega_1\alpha_1, \omega_1\alpha_2, \cdots, \omega_2\beta_1, \omega_2\beta_2, \cdots\}$ *is a free basis.*

Proof. Let H be a free group with free basis $\{x_1, x_2, \cdots, y_1, y_2, \cdots\}$ such that the functions ψ_1 and ψ_2 defined by

$$\psi_1\alpha_j = x_j, \qquad j = 1, 2, \cdots,$$
$$\psi_2\beta_k = y_k, \qquad k = 1, 2, \cdots,$$

are one-one correspondences between $\{\alpha_1, \alpha_2, \cdots\}$ and $\{x_1, x_2, \cdots\}$ and between $\{\beta_1, \beta_2, \cdots\}$ and $\{y_1, y_2, \cdots\}$, respectively. Since G_1 and G_2 are free, these correspondences extend to homomorphisms

$$\psi_i\colon G_i \to H, \qquad i = 1, 2.$$

Since G_0 is trivial, there is a corresponding trivial homomorphism $\psi_0\colon G_0 \to H$, and, in addition $\psi_0 = \psi_1\theta_1 = \psi_2\theta_2$. By the van Kampen theorem, there exists a homomorphism $\lambda\colon G \to H$ such that $\psi_i = \lambda\omega_i$, $i = 0, 1, 2$. Consequently

$$\lambda\omega_1\alpha_j = \psi_1\alpha_j = x_j, \qquad j = 1, 2, \cdots,$$
$$\lambda\omega_2\beta_k = \psi_2\beta_k = y_k, \qquad k = 1, 2, \cdots.$$

Since H is free, there exists a homomorphism $\mu\colon H \to G$ defined by

$$\mu x_j = \omega_1\alpha_j, \qquad j = 1, 2, \cdots,$$
$$\mu y_k = \omega_2\beta_k, \qquad k = 1, 2, \cdots.$$

Obviously, both compositions $\lambda\mu$ and $\mu\lambda$ are identity mappings. Hence, both are isomorphisms onto and inverses of each other, and the proof is complete.

(3.4) *If X_2 is simply-connected, then the homomorphism ω_1 is onto. Furthermore, if* $\{\alpha_1, \alpha_2, \cdots\}$ *generates G_0, then the kernel of ω_1 is the consequence of* $\{\theta_1\alpha_1, \theta_1\alpha_2, \cdots\}$.

Proof. Since G_2 is trivial, the image group $\omega_1 G_1$ generates G. No group can be generated by a proper subgroup; so $\omega_1 G_1 = G$. Turning to the second assertion, we observe that

$$\omega_1\theta_1\alpha_j = \omega_2\theta_2\alpha_j = 1, \qquad j = 1, 2, \cdots.$$

Hence, the consequence of $\{\theta_1\alpha_1, \theta_1\alpha_2, \cdots\}$ is contained in the kernel of ω_1. Conversely, consider an arbitrary element β in the kernel of ω_1 and the canonical homomorphism $\psi_1\colon G_1 \to H$, where H is the quotient group $G_1/$ (consequence of $\{\theta_1\alpha_1, \theta_1\alpha_2, \cdots\}$). The composition $\psi_1\theta_1$, which we denote by ψ_0, is, of course, trivial. Denoting the trivial homomorphism of G_2 into H

by ψ_2, we obtain the consistency relations $\psi_0 = \psi_1\theta_1 = \psi_2\theta_2$. By the van Kampen theorem, there exists a homomorphism $\lambda\colon G \to H$ such that $\psi_1 = \lambda\omega_1$. Therefore

$$\psi_1\beta = \lambda\omega_1\beta = 1,$$

and it follows that β is a consequence of $\{\theta_1\alpha_1, \theta_1\alpha_2, \cdots\}$. This completes the proof.

We are now in a position to calculate the fundamental groups of some interesting topological spaces. One drawback, however, will be the condition imposed on X_1 and X_2 by the van Kampen theorem that they be open sets. Most of the common examples don't come that way naturally, and we shall have to do a little prodding. Actually, under certain conditions, the van Kampen theorem holds for closed sets X_1, X_2 and could be applied to all our examples directly.[6] However, without introducing a good bit more terminology, it is difficult to describe the proper generalization succinctly. What has to be done in each case will become clear as we proceed.

(i) *The n-leafed rose.* This space, denoted by $C_{(n)}$, is the union of n topological circles $X_1, \cdots, X_n$ joined at a point p and otherwise disjoint (cf. Figure 22). The fundamental group of $C_{(n)}$ is free of rank n. More fully, if x_i

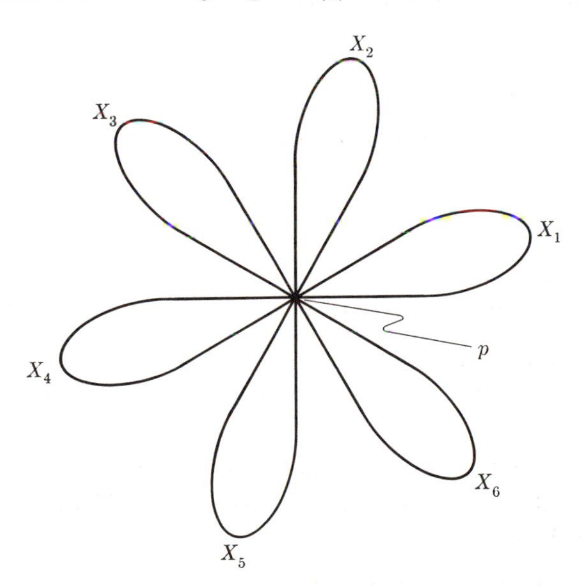

Figure 22

[6] E. R. van Kampen, "On the Connection between the Fundamental Groups of Some Related Spaces," *American Journal of Mathematics*, Vol. 55 (1933), pp. 261–267; P. Olum, "Nonabelian Cohomology and van Kampen's Theorem," *Ann. of Math.*, Vol. 68 (1958), pp. 658–668.

is a generator of the infinite cyclic group $\pi(X_i)$, and $\omega_i\colon \pi(X_i) \to \pi(C_{(n)})$ is induced by inclusion, then

$$\pi(C_{(n)}) = |\omega_1 x_1, \cdots, \omega_n x_n \ : \ |.$$

The proof is by induction on n. The space $C_{(1)}$ is a circle, whose group has been shown in Chapter II to be infinite cyclic, i.e., free of rank 1. Consider

$$C_{(n+1)} = C_{(n)} \cup X_{n+1},$$
$$\{p\} = C_{(n)} \cap X_{n+1}.$$

Except for the fact that $C_{(n)}$, X_{n+1}, and $\{p\}$ are not open subsets of $C_{(n+1)}$, the desired conclusion follows immediately from (3.3). To get around the difficulty, consider an open neighborhood N of p in $C_{(n+1)}$ consisting of p and the union of $2(n+1)$ disjoint, open arcs each of which has p as one of its endpoints (cf. Figure 23). Then $C_{(n)}$, X_{n+1}, and $\{p\}$ are deformation retracts of $C_{(n)} \cup N$,

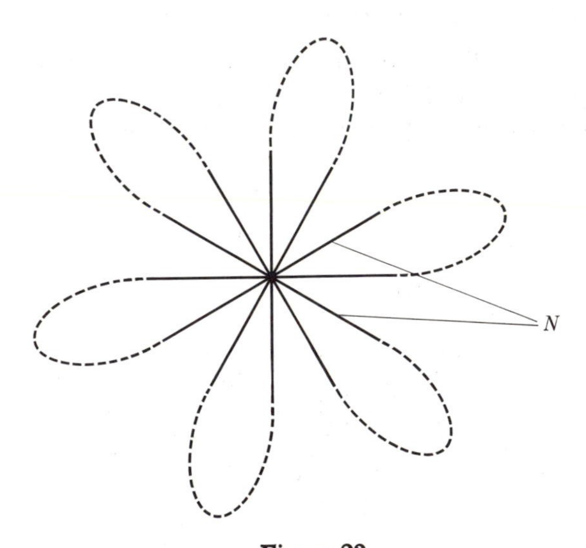

Figure 23

$X_{n+1} \cup N$, and $\{p\} \cup N$, respectively. The latter are open subsets of $C_{(n+1)}$, and (3.3) may be applied to complete the proof. In particular, for $n = 2$, we obtain the answer to the problem posed in the introduction to this chapter—that of computing the fundamental group of two circles joined at a point.

(ii) *The sphere.* Let X_0 be an open equatorial band dividing the sphere X into north and south polar caps. Set X_1 equal to the union of X_0 and the north polar cap and X_2, the union of X_0 and the south polar cap. Clearly, X_1, X_2, and $X_0 = X_1 \cap X_2$ are open, pathwise connected, and nonvoid in $X = X_1 \cup X_2$. Moreover, the spaces X_1 and X_2 are homeomorphic to convex discs and hence are simply-connected. By (3.2), it follows that the sphere is also simply-connected.

(iii) *The sphere with $n \geq 1$ holes.* By stretching one hole to an equator and projecting the result on a plane, one can see that the sphere with $n \geq 1$ holes, denoted by $S_{(n)}$, is topologically equivalent to a disc with $n - 1$ holes. If $n > 1$, the disc contains an $(n - 1)$-leafed rose as a deformation retract (cf. Figure 24). We conclude that $\pi(S_{(n)})$ is a free group of rank $n - 1$. (The trivial group is free of rank 0.)

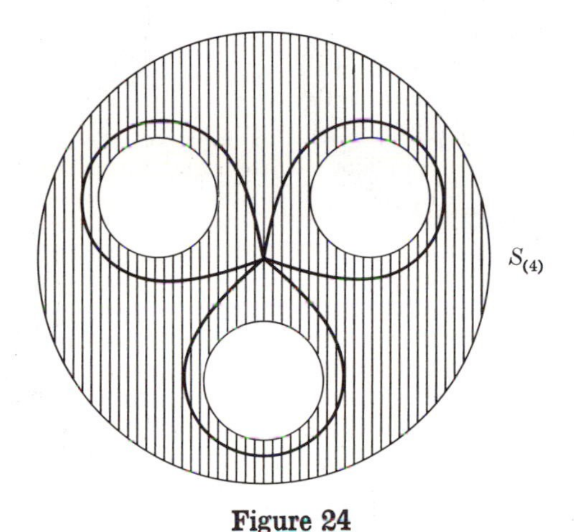

Figure 24

(iv) *The torus.* We shall exhibit the torus X as the union of two open subsets X_1 and X_2 such that X_2 is a disc and X_1 contains a 2-leafed rose as a deformation retract. The decomposition is pictured in Figure 25. The subspace X_1 is the torus minus a closed disc (or hole) D, and X_2 is an open disc of X which contains D. The intersection $X_0 = X_1 \cap X_2$ is an open annulus, and its fundamental group is therefore infinite cyclic. That X_1 is of the

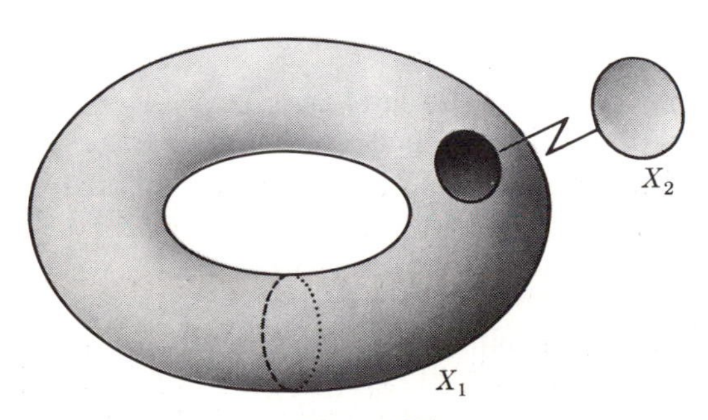

Figure 25

homotopy type of a 2-leafed rose can be seen by stretching the hole D as indicated in Figure 26. Hence, $\pi(X_1)$ is free of rank 2. A generator of $\pi(X_0)$ is represented by a path c running around the edge of X_1 and X_2. From Figure 26 it is clear that such a path is equivalent in X_1 to one running first

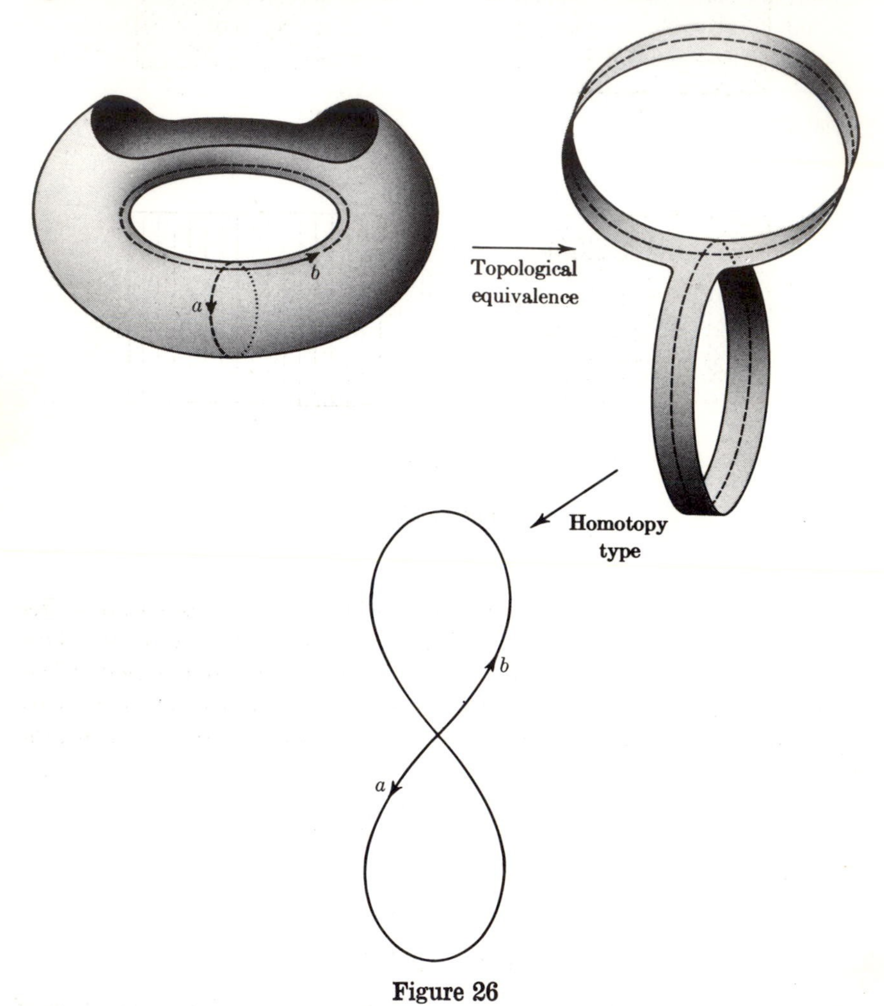

Figure 26

around a, then b, then a in the opposite direction, and finally around b in the opposite direction. Hence

$$[c] = [a][b][a]^{-1}[b]^{-1},$$

where the brackets indicate equivalence classes in X_1. Another good way to visualize this relation is to cut the torus along a and b and flatten it out as in Figure 27. The subsets X_1 and X_2 are shown as the shaded regions, and it is easy to read the above relation from the third picture of Figure 27. To

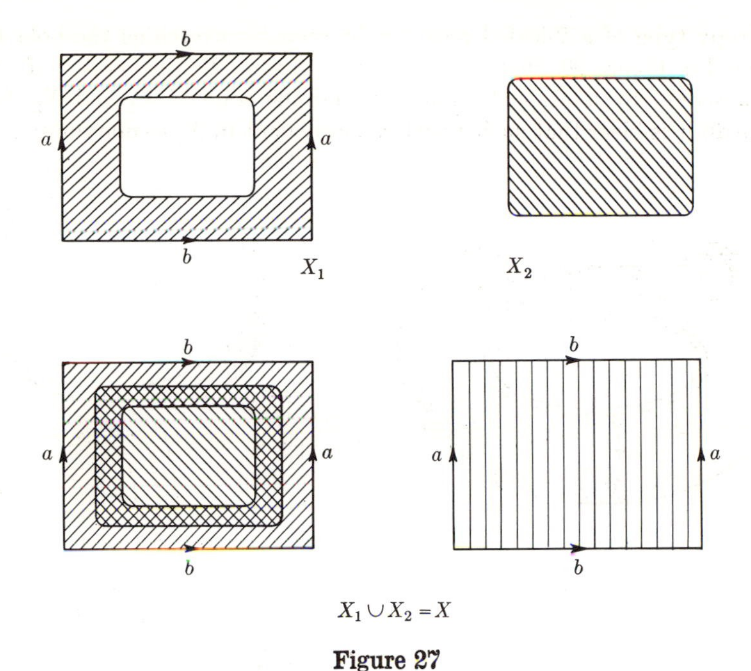

Figure 27

summarize: $\pi(X_1)$ is a free group and $x = [a]$ and $y = [b]$ constitute a free basis, X_2 is simply-connected, and $\pi(X_0)$ is generated by an element whose image in $\pi(X_1)$ under the homomorphism induced by inclusion is $xyx^{-1}y^{-1}$. It follows from (3.4) that the homomorphism $\omega_1 \colon \pi(X_1) \to \pi(X)$ induced by inclusion is onto and that its kernel is the consequence of $xyx^{-1}y^{-1}$. Thus, the group of the torus has the presentation

$$(x,y \ : \ xyx^{-1}y^{-1})$$

or, in the language of relations,

$$(x,y \ : \ xy = yx).$$

This group is the free abelian group of rank 2.

The final objective of this chapter is to derive from (3.1) a formulation of the van Kampen theorem in terms of group presentations. The spaces X and X_i, $i = 0, 1, 2$, are assumed to satisfy all the conditions imposed in the paragraph preceding (3.1), and the notation for the various fundamental groups and homomorphisms induced by inclusion is also the same. In addition, we assume given group presentations

$$\begin{aligned} G_1 &= | \, \mathbf{x} \ : \ \mathbf{r} \, |_{\phi_1}, \\ G_2 &= | \, \mathbf{y} \ : \ \mathbf{s} \, |_{\phi_2}, \qquad (3.5) \\ G_0 &= | \, \mathbf{z} \ : \ \mathbf{t} \, |_{\phi_0}. \end{aligned}$$

The problem is to find a group presentation of G, and the solution is stated in (3.6).

We denote by F_1, F_2, and F_0 the free groups which are the domains of ϕ_1, ϕ_2, and ϕ_0, respectively. It is convenient to assume that $\mathbf{x}$ and $\mathbf{y}$ are disjoint and that their union is a free basis of a free group F. Thus, F_1 and F_2 are subgroups of F. There exists a unique homomorphism ϕ: $F \to G$ such that

$$\phi \mid F_i = \omega_i \phi_i, \qquad i = 1, 2.$$

Notice that $\phi\mathbf{r} = \phi\mathbf{s} = 1$. Inasmuch as F_0 is free, the mappings θ_1 and θ_2 can be lifted to the free groups, i.e., there exist homomorphisms $\bar{\theta}_i$: $F_0 \to F_i$, $i = 1, 2$, so that

$$\begin{array}{ccccc} F_1 & \xleftarrow{\bar{\theta}_1} & F_0 & \xrightarrow{\bar{\theta}_2} & F_2 \\ \downarrow{\scriptstyle \phi_1} & & \downarrow{\scriptstyle \phi_0} & & \downarrow{\scriptstyle \phi_2} \\ G_1 & \xleftarrow[\theta_1]{} & G_0 & \xrightarrow[\theta_2]{} & G_2 \end{array}$$

is a consistent diagram. Where $\mathbf{z} = \{z_1, z_2, \cdots\}$, consider the set of all elements $\bar{\theta}_1 z_k \bar{\theta}_2 z_k^{-1}$, $k = 1, 2, \cdots$, in F. Clearly,

$$\begin{aligned} \phi(\bar{\theta}_1 z_k \bar{\theta}_2 z_k^{-1}) &= (\omega_1 \phi_1 \bar{\theta}_1 z_k)(\omega_2 \phi_2 \bar{\theta}_2 z_k^{-1}) \\ &= (\omega_1 \theta_1 \phi_0 z_k)(\omega_2 \theta_2 \phi_0 z_k^{-1}) \\ &= \omega_0 \phi_0 (z_k z_k^{-1}) = 1. \end{aligned}$$

Thus, the consequence of $\mathbf{r} \cup \mathbf{s} \cup \{\bar{\theta}_1 z_k \bar{\theta}_2 z_k^{-1}\}$ is contained in the kernel of ϕ. We contend that the converse is also true. To prove it, consider an arbitrary homomorphism ψ: $F \to H$ which maps $\mathbf{r} \cup \mathbf{s} \cup \{\bar{\theta}_1 z_k \bar{\theta}_2 z_k^{-1}\}$ onto 1. It is then obvious that there exist homomorphisms ψ_i: $G_i \to H$, $i = 1, 2$, so that

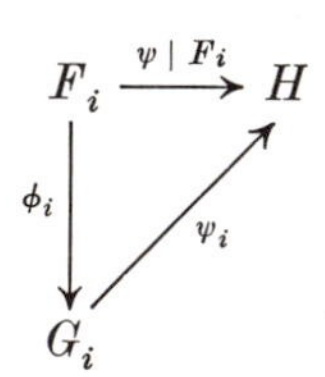

is consistent. Furthermore, since $\psi\bar{\theta}_1 z_k = \psi\bar{\theta}_2 z_k$, $k = 1, 2, \cdots$, it follows by diagram chasing that

$$\psi_1 \theta_1 \phi_0 z_k = \psi_2 \theta_2 \phi_0 z_k, \qquad k = 1, 2, \cdots.$$

Since the elements $\phi_0 z_1, \phi_0 z_2, \cdots$ generate G_0, we may conclude that the homomorphism ψ_0 is well-defined by

$$\psi_0 = \psi_1 \theta_1 = \psi_2 \theta_2.$$

By the van Kampen theorem (3.1), there exists a homomorphism λ: $G \to H$

such that $\psi_i = \lambda\omega_i$, $i = 0, 1, 2$. Consider, finally, an arbitrary element $u_i \in F_i$, $i = 1, 2$. Then,

$$\psi u_i = \psi_i\phi_i u_i = \lambda\omega_i\phi_i u_i = \lambda\phi u_i,$$

and hence

$$\psi = \lambda\phi.$$

It follows that anything in the kernel of ϕ is also in the kernel of ψ, and our contention is proved. Thus, the kernel of ϕ equals the consequence of

$$\mathbf{r} \cup \mathbf{s} \cup \{\bar{\theta}_1 z_k \bar{\theta}_2 z_k^{-1}\},$$

and we have proved

(3.6) Alternative Formulation of the van Kampen Theorem. *If the groups G_1, G_2, and G_0 have the presentations* (3.5), *then*

$$G = | \mathbf{x},\mathbf{y} : \mathbf{r},\mathbf{s}, \{\bar{\theta}_1 z_k \bar{\theta}_2 z_k^{-1}\} |_\phi.$$

EXERCISES

1. Consider the closed circular disc D with center p and boundary $\dot{D}$ as shown in Figure 17. We have seen that the complement in the plane R^2 of the open disc $\mathring{D} = D - \dot{D}$ is a deformation retract of the punctured plane $R^2 - p$. Prove that the complement $R^2 - D$ is not a deformation retract of $R^2 - p$, but that $R^2 - D$ and $R^2 - p$ are of the same homotopy type.

2. Prove that if X is a deformation retract of Y, and Y is a deformation retract of Z, then X is a deformation retract of Z.

3. Find a presentation for the fundamental group of (a) a Klein bottle, (b) a double torus.

4. Prove that the three spaces pictured in Figure 16 belong to the same homotopy type but to distinct topological types.

5. What is the fundamental group of the complement $R^3 - X$ for (a) X = circle; (b) X = union of two separated circles; (c) X = union of two simply linked circles?

CHAPTER VI

Presentation of a Knot Group

Introduction. In this chapter we return to knot theory. The major objective here is the description and verification of a procedure for deriving from any polygonal knot K in regular position two presentations of the group of K, which are called respectively the over and under presentations. The classical Wirtinger presentation is obtained as a special case of the over presentation. In a later section we calculate over presentations of the groups of four separate knots explicitly, and the final section contains a proof of the existence of nontrivial knots, in that it is shown that the clover-leaf knot can not be untied. The fact that our basic description in this chapter is concerned with a *pair* of group presentations represents a concession to later theory. It is of no significance at this stage. One presentation is plenty, and, for this reason, Section 4 is limited to examples of over presentations. The existence of a pair of over and under presentations is the basis for a duality theory which will be exploited in Chapter IX to prove one of the important theorems.

If K is any knot in 3-dimensional space R^3 and p_0 is any point in $R^3 - K$, then the fundamental group $\pi(R^3 - K,p_0)$ is called *the group of* K. Since $R^3 - K$ is connected, different choices of basepoint yield isomorphic groups. For this reason, it is common practice to omit explicit reference to the basepoint p_0 and speak simply of *the group* $\pi(R^3 - K)$ *of the knot* K. Nevertheless, the precise meaning of "the group of K" is always "the group $\pi(R^3 - K, p_0)$ for some basepoint p_0." It will be clear that the particular over and under presentations obtained from a given knot in regular position depend not only on the knot but also on a number of arbitrary choices. Hence, the terminology "the over and under presentations of the knot K" exemplifies to an even greater degree the same abuse of language as the phrase "the fundamental group of the space X." A knot in regular position has many pairs of over and under presentations. All of these will be seen to be of the same type.

1. The over and under presentations. Let K be a polygonal knot in regular position and $\mathscr{P}$ the projection $\mathscr{P}(x,y,z) = (x,y,0)$ (cf. Chapter I, Section 3). For some positive integer n, we select a subset Q of K containing exactly $2n$ points no one of which is either an overcrossing or an undercrossing. These divide K into two classes of closed, connected segmented arcs, the *overpasses* and the *underpasses*, which alternate around the knot, i.e., each point in Q belongs to one overpass and one underpass. The subdivision is to be chosen so that no overpass contains an undercrossing and no underpass contains an overcrossing. The construction can, of course, be done in many different ways although ordinarily we would want n as small as possible. We denote the overpasses by $A_1, \cdots, A_n$ and their union $A_1 \cup \cdots \cup A_n$ by A, the underpasses by $B_1, \cdots, B_n$ and their union $B_1 \cup \cdots \cup B_n$ by B. The ordering is arbitrary. It should be obvious that there exists a semi-linear homeomorphism[1] of R^3 onto itself which displaces points vertically, i.e., parallel to the z-axis, such that the image of $A - Q$ lies above the xy-plane R^2 and that of $B - Q$ lies below R^2. Since K and its image under this homeomorphism are equivalent polygonal knots, we make the simplifying assumption that K is in the image position to begin with. It follows that $Q \subset R^2$.

Each presentation is made with respect to an orientation of K and of R^3. Accordingly, one of the two directions along the knot is chosen as positive, i.e., we draw an arrow on K. In R^3 we shall consistently refer orientations to a left-handed screw. Two basepoints are selected: one, p_0, lying above the knot and the other, p_0', below. For later convenience, we shall assume that $p_0 = (0,0,z_0)$ and $p_0' = (0,0,-z_0)$ for some positive z_0. (Then $(x,y,z) \in K$ implies that $-z_0 < z < z_0$.) Thus, a rotation of 180° about the x-axis carries one basepoint onto the other. Finally, we choose a point $q_0 \in R^2 - \mathscr{P}K$.

Let us call a path a in R^2 *simple* if it satisfies the following three conditions: It is polygonal, neither initial nor terminal point belongs to $\mathscr{P}K$, and a intersects $\mathscr{P}K$ in only a finite number of points, no one of which is a vertex of a or a vertex of $\mathscr{P}K$. Let $F(\mathbf{x})$ be an arbitrary free group freely generated by $\mathbf{x} = (x_1, \cdots, x_n)$. To each simple path a in $R^2 - \mathscr{P}B$ we assign an element $a^\#$ in $F(\mathbf{x})$ defined as follows:

$$a^\# = x_{i_1}^{\epsilon_1} \cdots x_{i_l}^{\epsilon_l},$$

where the projected overpasses crossed by a are, in order, $\mathscr{P}A_{i_1}, \cdots, \mathscr{P}A_{i_l}$, and where $\epsilon_k = 1$ or -1 according as a crosses under A_{i_k} from left to right or from right to left (in other words, according as A_{i_k} and the path a form a left-handed or a right-handed screw). The assignment $a \rightarrow a^\#$, illustrated in Figure 28, is clearly product preserving,

$$(a_1 \cdot a_2)^\# = a_1^\# a_2^\#.$$

It is not, however, necessarily a mapping onto $F(\mathbf{x})$. For any point $p \in R^2$, let $\bar{p}$ be the path which runs linearly from p_0 parallel to R^2 to a point directly over p and thence linearly down to p. For any path a in R^2, we set

$$*a = \overline{a(0)} \cdot a \cdot \overline{a(\| a \|)}^{-1}.$$

The group $F(\mathbf{x})$ is to be the free group of the over presentation. A homomorphism ϕ: $F(\mathbf{x}) \rightarrow \pi(R^3 - K, p_0)$ is defined as follows: Let a_j be a simple

[1] A mapping $R^3 \rightarrow R^3$ is *semi-linear* if its restriction to every compact straight-line segment is linear at all but a finite number of points. Thus, polygons go into polygons.

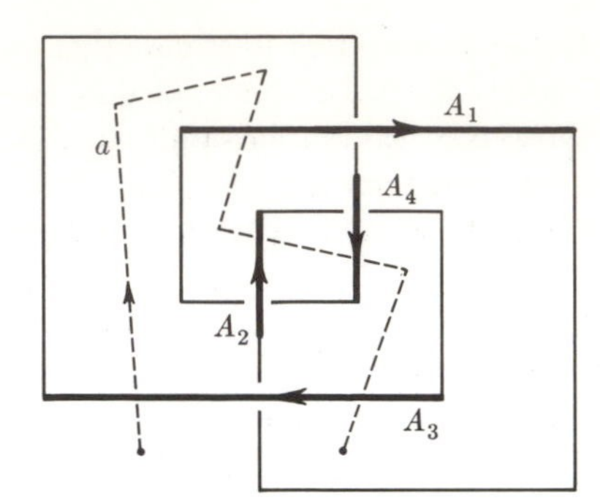

Figure 28. $a^\# = x_3x_1x_2x_4^{-1}x_3^{-1}$

path in $R^2 - \mathscr{P}B$ such that $a_j^\# = x_j, j = 1, \cdots, n$. We define

$$\phi x_j = [*a_j], \qquad j = 1, \cdots, n,$$

where the square brackets indicate the equivalence class in $R^3 - K$ of the p_0-based loop $*a_j$. It is obvious that ϕx_j is independent of the particular choice of representative path a_j. The homomorphism ϕ is the unique extension to the entire group $F(\mathbf{x})$ of this assignment on the generators, $x_1, \cdots, x_n$. It follows that

$$\phi a^\# = [*a],$$

for any path a in $R^2 - \mathscr{P}B$. It is our contention that the homomorphism ϕ is onto or, in other words, that $\phi x_1, \cdots, \phi x_n$ generate $\pi(R^3 - K, p_0)$. The proof is deferred until the next section; but it should be pointed out that the result is a very natural one. As suggested by Figure 29, it is geometrically almost obvious that every p_0-based loop in $R^3 - K$ is equivalent to a product

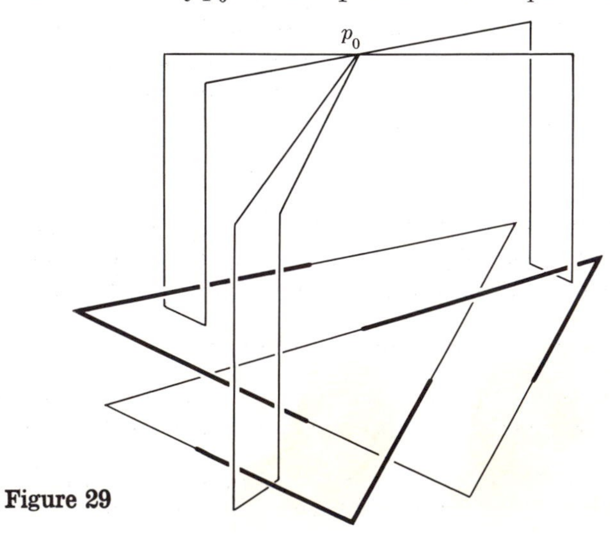

Figure 29

of the loops $*a_j$, $j = 1, \cdots, n$, and their inverses. It is hard to see what could go wrong.

The generators for the under presentation are chosen in an entirely analogous manner: Let $F(y)$ be a free group freely generated by $\mathbf{y} = (y_1, \cdots, y_n)$. To each simple path b in $R^2 - \mathscr{P}A$ we assign $b^\flat$ in $F(\mathbf{y})$ defined by

$$b^\flat = y_{j_1}{}^{\delta_1} \cdots y_{j_m}{}^{\delta_m},$$

where the projected underpasses crossed by b are in order $\mathscr{P}B_{j_1}, \cdots, \mathscr{P}B_{j_m}$, and $\delta_k = 1$ or -1 according as b crosses over B_{j_k} from left to right or from right to left (i.e., according as B_{j_k} and b form a right-handed or a left-handed screw). The assignment $b \to b^\flat$ is illustrated in Figure 30.

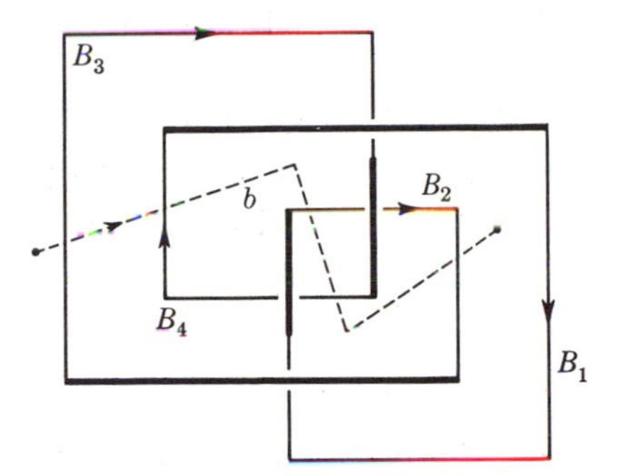

Figure 30 $b^\flat = y_3y_4y_2y_4^{-1}y_2^{-1}$

A homomorphism $\phi'\colon F(\mathbf{y}) \to \pi(R^3 - K, p_0')$ is defined in the same way: Where $\mathscr{R}$ is the reflection $\mathscr{R}(x,y,z) = (x,y,-z)$ and a is any path in R^2, we set $_*a = \mathscr{R}{*a}$. Let b_i be a simple path in $R^2 - \mathscr{P}A$ with $b_i{}^\flat = y_i$, $i = 1, \cdots, n$. Then,

$$\phi' y_i = [{}_*b_i], \qquad i = 1, \cdots, n,$$

where the square brackets indicate the equivalence class in $\pi(R^3 - K, p_0')$. Again, we contend that the extension of this assignment on the generators is a homomorphism onto.

The images $\mathscr{P}B_i$, $i = 1, \cdots, n$ of the underpasses are disjoint segmented arcs. Hence we may select disjoint, simply-connected, open sets $V_1, \cdots, V_n$ in R^2 such that $\mathscr{P}B_i \subset V_i$, $i = 1, \cdots, n$, and such that their boundaries are the disjoint images of simple loops $v_1, \cdots, v_n$ which run counterclockwise (from above) around $V_1, \cdots, V_n$, respectively. Similarly, we choose simply-connected, disjoint, open sets $U_1, \cdots, U_n$ in R^2 such that $\mathscr{P}A_j \subset U_j$, $j = 1, \cdots, n$, with boundaries that are the disjoint images of simple loops $u_1, \cdots, u_n$, which run clockwise (from above) around $U_1, \cdots, U_n$, respectively. We also insist that the previously chosen point q_0 lie outside the closures of all the regions V_i, U_j, $i, j = 1, \cdots, n$. Next, we select a set of simple paths $c_1, \cdots, c_n$ such that each c_i has initial point q_0 and terminal point $v_i(0)$, and $c_i(t) \in R^2 - \bigcup_{k=1}^{n} \bar{V}_k$ unless $t = \| c_i \|$. ($\bar{V}_k$ is the closure of V_k.) Similarly, we choose a set of simple paths $d_1, \cdots, d_n$ such that $d_j(0) = q_0$, and $d_j(\| d_j \|) = u_j(0)$, and $d_j(t) \in R^2 - \bigcup_{k=1}^{n} \bar{U}_k$ unless $t = \| d_j \|$. These paths may, of course, be chosen in several ways. Examples that illustrate these regions and paths are shown in Figures 32, 33, and 34.

We are now in a position to describe the two presentations. The *over presentation* of $\pi(R^3 - K, p_0)$ is

$$(1.1) \qquad (x_1, \cdots, x_n : r_1, \cdots, r_n)_\phi,$$

where $r_i = (c_i \cdot v_i \cdot c_i^{-1})^\#$, $i = 1, \cdots, n$. The corresponding *under presentation* of $\pi(R^3 - K, p_0')$ is

$$(1.2) \qquad (y_1, \cdots, y_n : s_1, \cdots, s_n)_{\phi'},$$

where $s_j = (d_j \cdot u_j \cdot d_j^{-1})^\flat$, $j = 1, \cdots, n$. The validity of the equations $\phi r_i = 1$ and $\phi' s_j = 1$, $i, j = 1, \cdots, n$, is geometrically easy to see. We have

$$\phi r_i = [{}^*(c_i \cdot v_i \cdot c_i^{-1})], \qquad i = 1, \cdots, n,$$

and

$$\phi' s_j = [{}_*(d_j \cdot u_j \cdot d_j^{-1})], \qquad j = 1, \cdots, n.$$

The contraction of a typical loop ${}^*(c_i \cdot v_i \cdot c_i^{-1})$ by sliding it below B_i is illustrated in Figure 31. The analogous picture can, of course, be drawn for the under presentation. Incidentally, we do not claim that it is obvious that the relators $r_1, \cdots, r_n$ and $s_1, \cdots, s_n$ constitute defining sets.

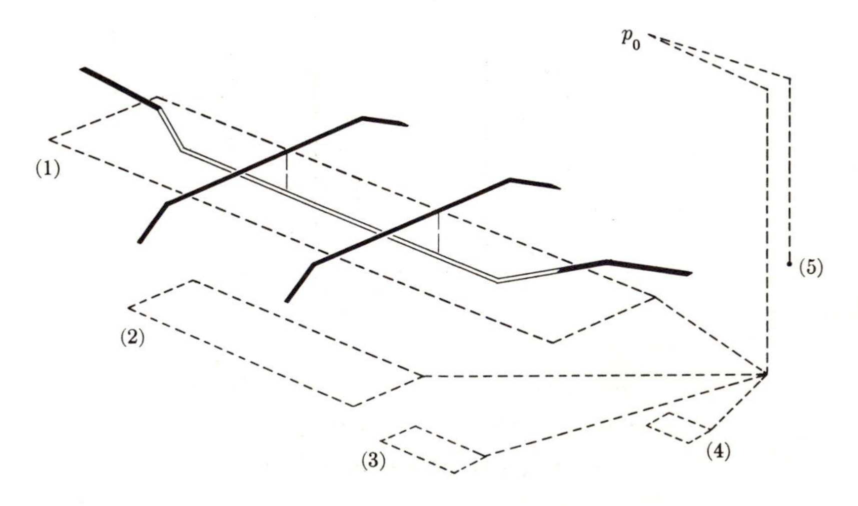

Figure 31

The description of the over and under presentations is almost as complicated as the subsequent proofs. We suggest that the reader carry out the construction in some examples. A fairly complicated sample computation is given below in Figures 32 and 33. Overpasses are drawn in heavy lines and underpasses in light. The paths c_i, d_j, v_i, u_j are drawn with dotted lines. It is convenient to indicate the generators x_j and y_i with small arrows, as we have done. Additional examples of over presentations are derived in Section 4.

It remains to prove that the over and under presentations are what we claim they are, i.e., group presentations of $\pi(R^3 - K, p_0)$ and $\pi(R^3 - K, p_0')$, respectively. The proof is given in the next section. An important corollary will be the theorem.

(1.3) *In any over presentation* (1.1) [*under presentation* (1.2)], *any one of the relators* $r_1, \cdots, r_n$ $[s_1, \cdots, s_n]$ *is a consequence of the other* $n - 1$.

Thus, in either presentation, any one of the relators may be dropped. This fact is a substantial aid to computation. We shall also see that it has significant theoretical implications.

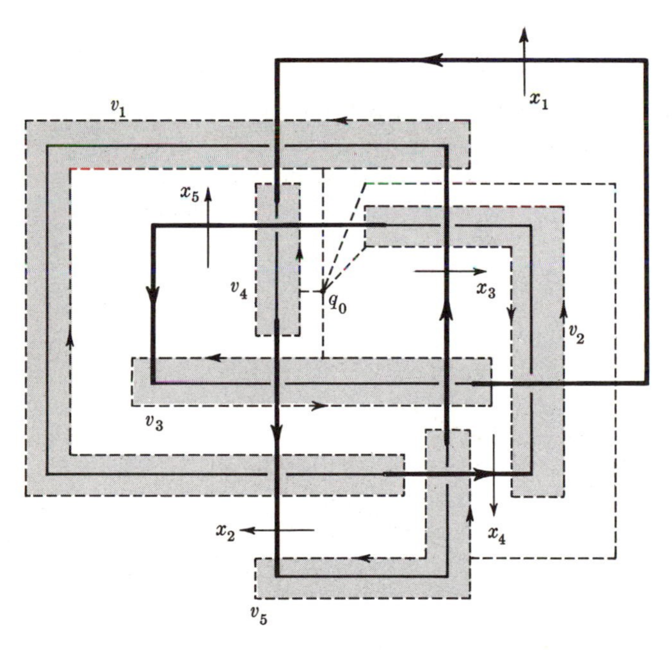

$r_1 = x_5(x_3 x_1 x_2^{-1} x_4^{-1} x_2 x_1^{-1}) x_5^{-1}$ $\quad$ $r_2 = x_3 x_1 x_4 x_1^{-1} x_3^{-1} x_5^{-1}$

$r_3 = x_2 x_5 x_2^{-1} x_3 x_1^{-1} x_3^{-1}$ $\quad$ $r_4 = x_5 x_1 x_5^{-1} x_2^{-1}$

$r_5 = x_5 x_3 x_1 (x_4^{-1} x_3^{-1} x_4 x_2) x_1^{-1} x_3^{-1} x_5^{-1}$

Figure 32

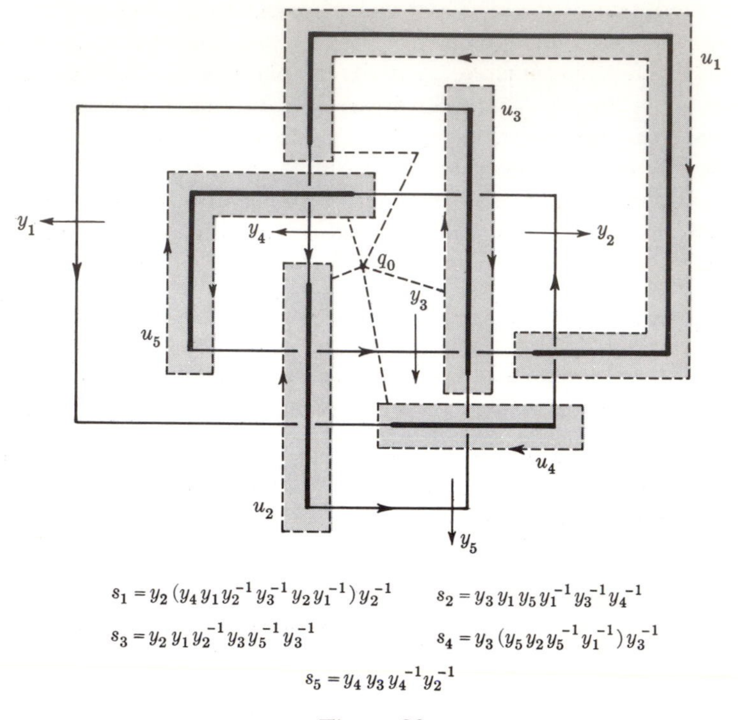

Figure 33

2. The over and under presentations, continued. In this section we shall prove that the over and under presentations (1.1) and (1.2) are in fact group presentations of $\pi(R^3 - K, p_0)$ and $\pi(R^3 - K, p_0')$, respectively. In so doing, we shall also obtain Theorem (1.3). We observe, first of all, that it is not necessary to give separate proofs for both presentations. An under presentation can be characterized in terms of an over presentation by simply reversing the orientation of K and interchanging the roles of up and down. More explicitly, let $h\colon\ R^3 \to R^3$ be the rotation of $180°$ about the x-axis defined by $h(x,y,z) = (x,-y,-z)$. Define $K' = hK$. The homeomorphism h induces the isomorphism $h_*\colon\ \pi(R^3 - K', p_0) \to \pi(R^3 - K, p_0')$. We take as the positive direction along K' the opposite of that induced by h from the orientation on K that was used in defining (1.1). Simply speaking, we turn K over and reverse the arrow. Let

$$(y_1, \cdots, y_n : s_1, \cdots, s_n)_\zeta$$

be any over presentation of $\pi(R^3 - K', p_0)$ constructed in the same way as (1.1). We assume that the jth overpass of K' is hB_j and that the ith relator is obtained by reading around $\mathscr{P}hA_i$ and, finally, that hq_0 is the common basepoint in R^2 of the loops determining the relators. Then,

$$| y_1, \cdots, y_n : s_1, \cdots, s_n |_{h_*\zeta} = \pi(R^3 - K, p_0')$$

obviously coincides with (1.2). To see that it is obvious, try an example. Figure 34 shows the over presentation obtained by rotating the knot of Figures 32 and 33 by 180° and reversing the arrow. As a result of these remarks, we shall restrict our attention to the over presentation in the remainder of this section.

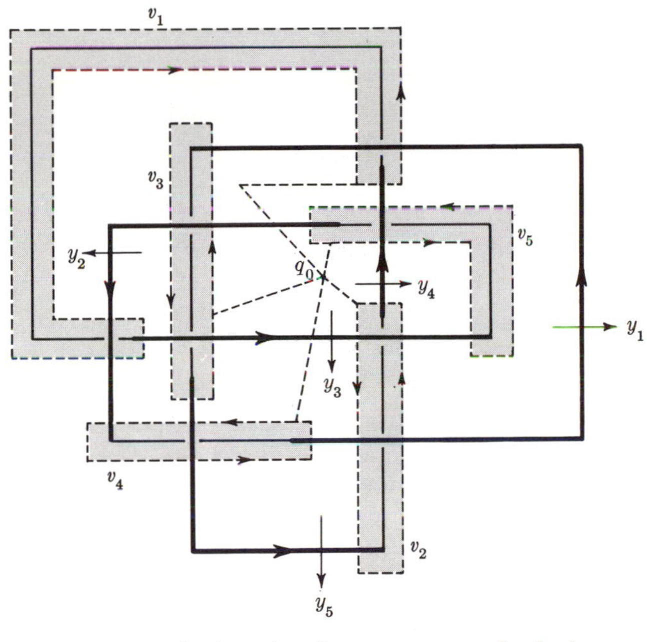

$$s_1 = y_2 (y_4 y_1 y_2^{-1} y_3^{-1} y_2 y_1^{-1}) y_2^{-1} \qquad s_2 = y_3 y_1 y_5 y_1^{-1} y_3^{-1} y_4^{-1}$$

$$s_3 = y_2 y_1 y_2^{-1} y_3 y_5^{-1} y_3^{-1} \qquad s_4 = y_3 (y_5 y_2 y_5^{-1} y_1^{-1}) y_3^{-1}$$

$$s_5 = y_4 y_3 y_4^{-1} y_2^{-1}$$

Figure 34

Consider a closed square S (boundary plus interior) parallel to the plane R^2, lying below K, and such that

$$\mathscr{P}K \cup \{q_0\} \cup \bigcup_{i=1}^{n} V_i \subset \mathscr{P}S. \tag{2.1}$$

For any subset L of K, we denote by L_* the union of K, S, and the set of all points (x,y,z) which lie between S and L, i.e., which satisfy $z_1 \leq z \leq z_2$, where $(x,y,z_1) \in S$, $(x,y,z_2) \in L$. For example, Q_* is the union of K, S, and the $2n$

vertical segments joining S to Q, the set of points of the original subdivision which separates the knot into overpasses and underpasses. The set K_* for a typical clover-leaf knot is illustrated in Figure 35.

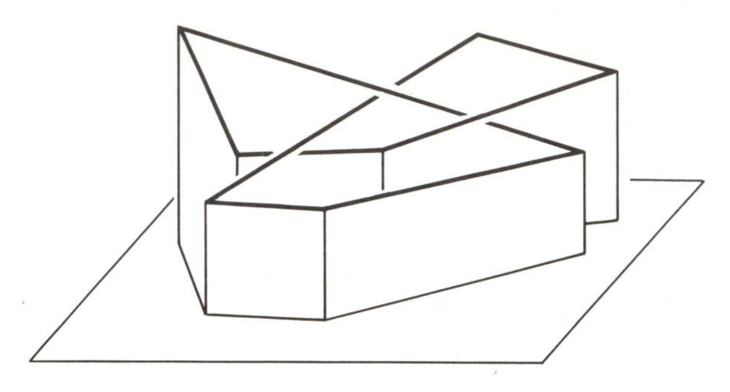

Figure 35

The derivation is a sequence of applications of the van Kampen theorem. The space $R^3 - K_*$ is first shown to be simply-connected. Next, the set $K_* - B_*$ is adjoined to $R^3 - K_*$, and it is proved that $\pi(R^3 - B_*, p_0)$ is a free group of rank n. A final application of the van Kampen theorem fills in $B_* - K$ and yields an over presentation of $\pi(R^3 - K, p_0)$.

(2.2) *The space* $R^3 - K_*$ *is simply-connected.*

Proof. The result is geometrically obvious (cf. Figure 35), but a formal proof using the van Kampen theorem can be given: Let X be the set of all points of $R^3 - K_*$ which do not lie below S. That is, if S' is the set of all points (x,y,z') such that $(x,y,z) \in S$ and $z' < z$, then

$$X = (R^3 - K_*) - S'.$$

It is not only intuitively apparent but also easy to prove that X is simply-connected. In fact, the basepoint $p_0 = (0,0,z_0)$ is a deformation retract of X. This fact is obtained from the following two deformations:

$$h_s(x,y,z) = (x, y, (1 - s)z + sz_0), \qquad 0 \leq s \leq 1,$$

$$k_s(x,y,z_0) = ((1 - s)x, (1 - s)y, z_0), \qquad 0 \leq s \leq 1,$$

The first, $\{h_s\}$, is a vertical deformation of X onto the horizontal plane containing p_0; the second, $\{k_s\}$, collapses this plane onto p_0 (cf. Exercise 2, Chapter V to justify composition of these deformations). Next, set Y equal to the simply-connected space consisting of all points lying below the horizontal plane that contains the square S. Clearly,

$$X \cup Y = R^3 - K_*.$$

The sets X, Y, and $X \cap Y$ are nonvoid, open, and pathwise connected in $X \cup Y$. It follows from the van Kampen theorem, or more specifically Corollary (3.2), Chapter V, that $X \cup Y$ is simply-connected, and the proof of (2.2) is complete.

The set $K_* - B_*$ is the disjoint union of n open topological discs $F_1, \cdots, F_n$, which we order so that F_j consists of those points lying between $A_j - Q$ and S, minus any points that happen also to lie on or below an underpass (cf. Figure 36). For each j, $j = 1, \cdots, n$, let a_j be a simple path in $R^2 - \mathscr{P}B$

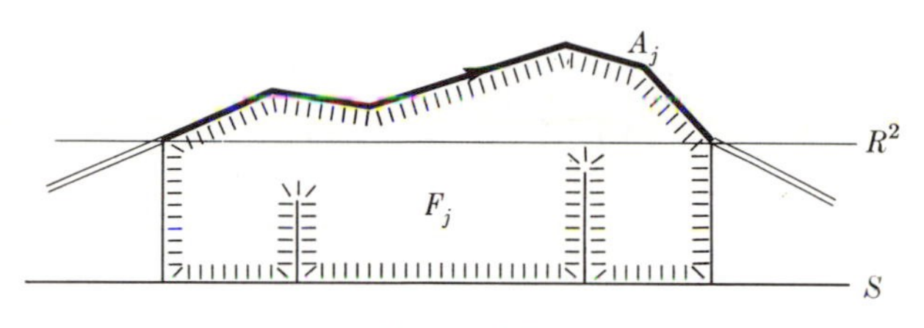

Figure 36

which crosses under A_j once, from left to right. The loop $*a_j$ intersects F_j once and is otherwise contained in $R^3 - K_*$. Let W_j be an open neighborhood of the union of F_j and the set of image points of $*a_j$ chosen so that:

(i) W_j is pathwise connected, and the group $\pi(W_j, p_0)$ is infinite cyclic and generated by the equivalence class of $*a_j$ in W_j.
(ii) $W_j \cap K_* = F_j$.
(iii) $W_j - K_*$ is simply-connected.

That these sets can be constructed is geometrically obvious. Each W_j is just the union of F_j and the image of $*a_j$—both slightly "thickened" to an open set in R^3. More explicitly, let $\epsilon > 0$ be so small that the open ϵ-neighborhood W_j' of the image of $*a_j$ satisfies $W_j' \cap K_* \subset F_j$. Set W_j'' equal to the set of all points whose distance from F_j is less than ϵ and which are closer to F_j than to $K_* - F_j$. Proving that $W_j = W_j' \cup W_j''$ satisfies the above conditions (i), (ii), and (iii) would be admittedly tedious, but presumably possible.

(2.3) *$R^3 - B_*$ is pathwise connected and $\pi(R^3 - B_*, p_0)$ is a free group. Furthermore, the set of equivalence classes $x_1, \cdots, x_n$ of the loops $*a_1, \cdots, *a_n$ is a free basis.*

Proof. Set

$$X_0 = R^3 - K_*,$$

$$X_j = X_{j-1} \cup W_j, \qquad j = 1, \cdots, n.$$

Since $W_j \cap K_* = F_j$,

$$(R^3 - K_*) \cup W_j = R^3 - (K_* - W_j) = R^3 - (K_* - F_j)$$
$$= (R^3 - K_*) \cup F_j,$$

and so

$$X_j = (R^3 - K_*) \cup \bigcup_{k=1}^{j} F_k, \qquad j = 1, \cdots, n.$$

In particular,

$$X_n = (R^3 - K_*) \cup (K_* - B_*)$$
$$= R^3 - B_*,$$

and so the sequence terminates in the right space. Furthermore, since

$$F_k \cap W_j = K_* \cap W_k \cap W_j = F_k \cap F_j = \emptyset, \qquad k \neq j,$$

it follows that

$$X_{j-1} \cap W_j = (R^3 - K_*) \cap W_j$$
$$= W_j - K_*,$$

and thus, by (iii), that $X_{j-1} \cap W_j$ is simply-connected. Lemma (2.2) states that X_0 is simply-connected, and we take as an inductive hypothesis that X_{j-1} is pathwise connected, $\pi(X_{j-1}, p_0)$ is a free group, and the set of equivalence classes of $*a_1, \cdots, *a_{j-1}$ in X_{j-1} is a free basis. Since X_{j-1}, W_j, and $X_{j-1} \cap W_j$ are pathwise connected, nonvoid, open subsets of $X_j = X_{j-1} \cup W_j$, it follows by (i) above and by (3.3) of Chapter V, that the set of equivalence classes in X_j of $*a_1, \cdots, *a_j$ is a free basis of $\pi(X_j, p_0)$. Induction completes the proof.

The elements $x_1, \cdots, x_n$ introduced in the preceding lemma are to be the generators of the over presentation. In other words, the group $\pi(R^3 - B_*, p_0)$ is to be taken as the free group of the presentation. Notice that, where

$$\text{(2.4)} \qquad \phi\colon \pi(R^3 - B_*, p_0) \rightarrow \pi(R^3 - K, p_0)$$

is the homomorphism induced by inclusion, the element ϕx_j, $j = 1, \cdots, n$, is the equivalence class in $R^3 - K$ of the loop $*a_j$. Thus, to within an isomorphism, ϕ is identical with the homomorphism denoted by the same symbol in Section 1. For any simple path a in $R^2 - \mathscr{P}B$, the element $a^\#$ will be understood to be a member of $\pi(R^3 - B_*, p_0)$. In fact, $a^\#$ is just the equivalence class in $R^3 - B_*$ of the loop $*a$. It is our final contention, which completes the derivation, that ϕ is onto and that its kernel is the consequence of any $n - 1$ of the relators $r_1, \cdots, r_n$ that occur in (1.1). Notice that this proposition includes Theorem (1.3).

A proof is obtained by the same technique that was used to determine the group $\pi(R^3 - B_*, p_0)$. By adjoining to the space whose group is known an appropriate open neighborhood of the set to be filled in (here, $B_* - K$), the unknown group structure is obtained by an application of the van Kampen theorem. To this end, consider a rectangular box T which contains the square S

in its interior and the knot K in its exterior. Topologically, T is just a sphere; but for convenience it will be assumed to be thin and flat with two faces parallel to S. Let W be an open neighborhood of $B_* - K$ chosen so that:

(iv) W is simply-connected and contains p_0.
(v) $W \cap K = \emptyset$.
(vi) $T - B_*$ is a deformation retract of $W - B_*$.

Such a set can be constructed in many ways. For instance, connect T to p_0 with a polygonal arc E which is disjoint to K_*. Let W' be the union of an ϵ-neighborhood of $T \cup E$ and the points inside T. Set W'' equal to the set of all points whose distance from B_* is less than ϵ and which are closer to $B_* - K$ than to K. For sufficiently small ϵ, the set $W' \cup W''$ may be taken as W.

The goal of this section is the following theorem.

(2.5) *The knot group of K has the over presentation*

$$\pi(R^3 - K, p_0) = |\, x_1, \cdots, x_n \,:\, r_1, \cdots, \hat{r}_k, \cdots, r_n \,|_\phi$$

where $\hat{r}_k$ indicates the deletion of the kth relator r_k and ϕ is the homomorphism (2.4).

Proof. We shall apply the van Kampen theorem to the groups of $R^3 - B_*$, W, and $(R^3 - B_*) \cap W$. We first observe that

$$\begin{aligned} B_* - W &= (K \cup (B_* - K)) - W \\ &= (K - W) \cup ((B_* - K) - W) = K. \end{aligned}$$

Hence,

$$(R^3 - B_*) \cup W = R^3 - (B_* - W) = R^3 - K,$$

and so this union is the space whose group we are after. Also,

$$(R^3 - B_*) \cap W = W - B_*$$

and $T - B_*$ belong to the same homotopy type. The space $T - B_*$ is a box with n knife cuts in the top; it is therefore of the homotopy type of a sphere with n holes and its fundamental group is free of rank $n - 1$ (cf. Example (iii), Chapter V, Section 3). In greater detail: Let h be the vertical projection of the plane R^2 upon the plane containing the top of T. It is a consequence of (2.1) that

$$\left(\{hq_0\} \cup \bigcup_{i=1}^{n} hV_i\right) \subset T.$$

Since the closures $\bar{V}_1, \cdots, \bar{V}_n$ of the regions $V_1, \cdots, V_n$ are pairwise disjoint, we may select a set of polygonal paths $e_1, \cdots, e_n$ in $T - \bigcup_{i=1}^n hV_i$ with the common initial point hq_0 and subject to the following restrictions: Each path e_i is an arc (i.e., a homeomorphism) and its terminal point is $h(v_i(0))$ (i.e., the image under h of the basepoint of the loop which bounds V_i).

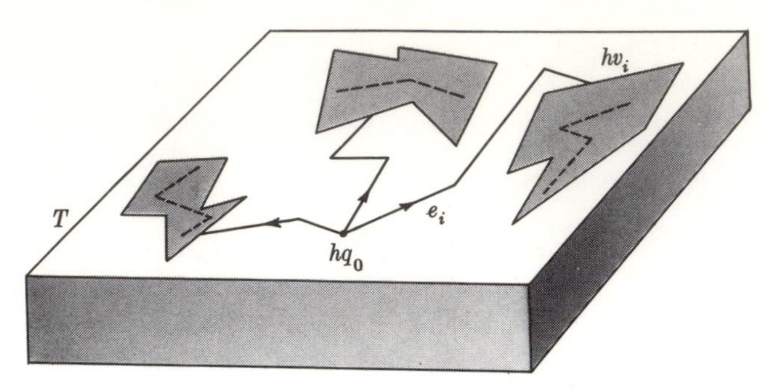

Figure 37

The images of $e_1, \cdots, e_n$ are, except for the point hq_0, pairwise disjoint and are contained, except for their terminal points, in $T - \bigcup_{i=1}^n h\bar{V}_i$ (cf. Figure 37).

Set

$$w_i = e_i \cdot hv_i \cdot e_i^{-1}, \qquad i = 1, \cdots, n.$$

Clearly, the union of the images of the loops $w_1, \cdots, w_n$ is of the homotopy type of an n-leafed rose (cf. Example (iii), Chapter V). Furthermore, the union of the images of any $n - 1$ of $w_1, \cdots, w_n$ is a deformation retract of $T - B_*$. Although a completely rigorous proof of this fact would admittedly be a nontrivial affair, the geometric idea is simple: First, widen all the knife cuts and push them back onto the curves hv_i. Then, choose one of the holes and starting from it, push the rest of the box onto the remaining $n - 1$ loops w_i. In any event, we conclude that $\pi(T - B_*, hq_0)$ is a free group and the equivalence classes of any $n - 1$ of $w_1, \cdots, w_n$ constitute a free basis. It follows that the same is true of the group $\pi(W - B_*, hq_0)$, and finally, where a is a path in $W - B_*$ which runs from p_0 to hq_0, *the set of equivalence classes of any* $n - 1$ *of* $a \cdot w_i \cdot a^{-1}$, $i = 1, \cdots, n$, *is a free basis of* $\pi(W - B_*, p_0)$.

It is then a direct consequence of Corollary (3.4), Chapter V, of the van Kampen theorem that the homomorphism ϕ is onto and that its kernel is the consequence of any $n - 1$ of the equivalence classes in $R^3 - B_*$ of the loops $a \cdot w_i \cdot a^{-1}$, $i = 1, \cdots, n$. Thus we obtain

$$\pi(R^3 - K, p_0) = | \, x_1, \cdots, x_n \; : \; [a \cdot w_i \cdot a^{-1}], \quad i = 1, \cdots, \hat{k}, \cdots, n \, |_\phi.$$

The proof is completed with the observation that, for each $i = 1, \cdots, n$, the element $[a \cdot w_i \cdot a^{-1}]$ is conjugate to $v_i^\#$ and, thence, to r_i. To prove this, consider a path b_i (analogous to and equivalent to $\overline{v_i(0)}$) which runs in a straight horizontal line from p_0 to the point directly over $v_i(0)$ and thence straight down to $hv_i(0)$. It is obvious (cf. Figure 38) that in $R^3 - B_*$ a valid equivalence is

$$*v_i \simeq b_i \cdot hv_i \cdot b_i^{-1}, \qquad i = 1, \cdots, n.$$

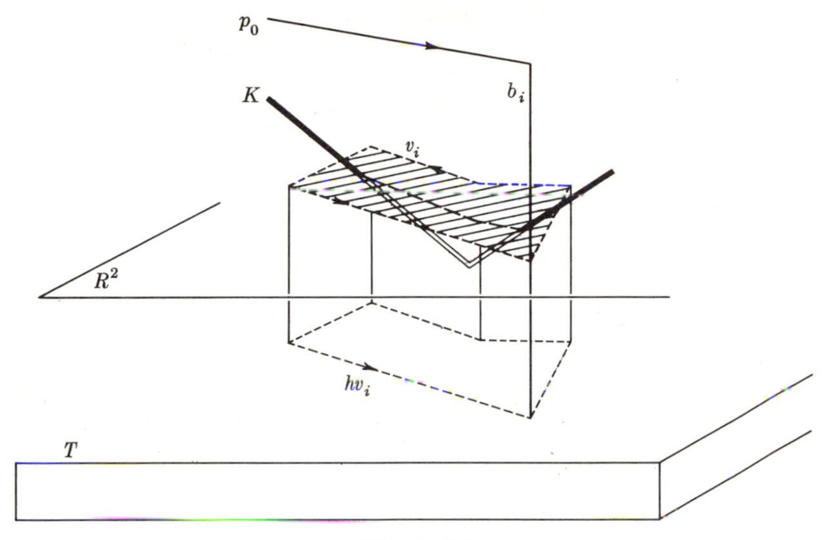

Figure 38

Consequently,

$$\begin{aligned} a \cdot w_i \cdot a^{-1} &= (a \cdot e_i) \cdot hv_i \cdot (a \cdot e_i)^{-1} \\ &\simeq (a \cdot e_i \cdot b_i^{-1}) \cdot b_i \cdot hv_i \cdot b_i^{-1} \cdot (a \cdot e_i \cdot b_i^{-1})^{-1} \\ &\simeq f_i \cdot {}^*v_i \cdot f_i^{-1}, \end{aligned}$$

where $f_i = (a \cdot e_i \cdot b_i^{-1})$. As a result,

$$[a \cdot w_i \cdot a^{-1}] = [f_i]v_i^{\#}[f_i]^{-1}, \qquad i = 1, \cdots, n.$$

Since $v_i^{\#}$ is in turn conjugate to r_i, the kernel of ϕ is the consequence of $r_1, \cdots, \hat{r}_k, \cdots, r_n$, and the proof of (2.5) is complete.

It is not unlikely that an intelligent reader, having worked his way through this section, will be unsatisfied. In the first place, the derivation is long and complicated. What is more unsettling, however, may be the feeling that, in spite of its length, it is still incomplete. At three places the existence of the complicated geometric construction essential to the argument is assumed without proof. In the interests of both economy and elegance, would it not be better simply to assume the desired conclusion and be done with it? The most honest answer is that there are degrees of the obvious. The first section of this chapter leaves two real questions unanswered: How do we really know that the elements $\phi x_1, \cdots, \phi x_n$ generate the knot group? And the harder one: Why do any $n - 1$ of $r_1, \cdots, r_n$ constitute a defining set of relators? In contrast, the assertions whose details we omitted are of a different sort. One wonders perhaps how to prove them in the best way, but not whether or not they are true. In spite of this preamble, the main point of this paragraph is not to present an apology. It is rather to call attention to the fact that, although these omitted details are intricate, they are not profound. The reason simply stems from our exclusive use of finite polygonal constructions. Everything can be chopped up into a finite number of triangles and tetrahedra, and these can be studied one at a time. It is tempting to try to eliminate these restrictions of linearity—if for no other reason than that it is unnatural to draw segmented knots. To some degree this can be done. However, the existence of wild knots whose groups are not finitely generated means that more is involved than simply convenience.

3. The Wirtinger presentation. An essential feature of any pair of corresponding over and under presentations is the common basepoint, denoted in Section 1 by q_0, of the $2n$ simple paths in R^2 whose images under $\sharp$ and $\flat$ constitute the relators of the two presentations. The necessity of having a common point will only become clear when the precise duality between over and under presentation is studied in Chapter IX. Notice, however, that whatever the reason may be, it has nothing to do with presentation type. The presentation

$$(3.1) \qquad (x_1, \cdots, x_n : v_1{}^{\sharp}, \cdots, v_n{}^{\sharp})_{\phi}$$

of $\pi(R^3 - K, p_0)$ is obtained from the over presentation (1.1) by n applications of Tietze operations **I** and **I**$'$. (We recall that v_i is the loop in $R^2 - \mathscr{P}B$ around the projected underpass $\mathscr{P}B_i$.) The main advantage of (3.1) over (1.1) is simply that there is less work in calculating it; one need not bother finding the elements $c_1{}^{\sharp}, \cdots, c_n{}^{\sharp}$. Therefore, we have used this modified form of the over presentation exclusively in the examples in the next section. Of course, it is also true of (3.1) that *any one of the relators* $v_1{}^{\sharp}, \cdots, v_n{}^{\sharp}$ *is a consequence of the remaining* $n - 1$ *of them and therefore may be omitted.*

A presentation (3.1) of the group of a given knot is called a *Wirtinger presentation* if each underpass contains exactly one undercrossing and each path v_i cuts the projected overpasses in just four places. These two conditions can always be imposed unless, of course, the knot has no undercrossings. That they are natural restrictions to make is evidenced by the fact that historically this presentation of a knot group was one of the first to be studied, and it is certainly the commonest one encountered in the literature. The presentations of the clover-leaf knot and of the figure-eight knot in the next section are examples. An attractive feature of the Wirtinger presentation is that the relators are particularly simple: written as a relation, each one is of the form $x_k = x_i^{\epsilon} x_l x_i^{-\epsilon}$, $\epsilon = \pm 1$ (cf. Figure 39). Notice, however, that unless the knot projection is alternating (i.e., as one traverses the knot, undercrossings and overcrossings alternate), the Wirtinger presentation is not the most economical in the number of generators and relators which might be obtained.

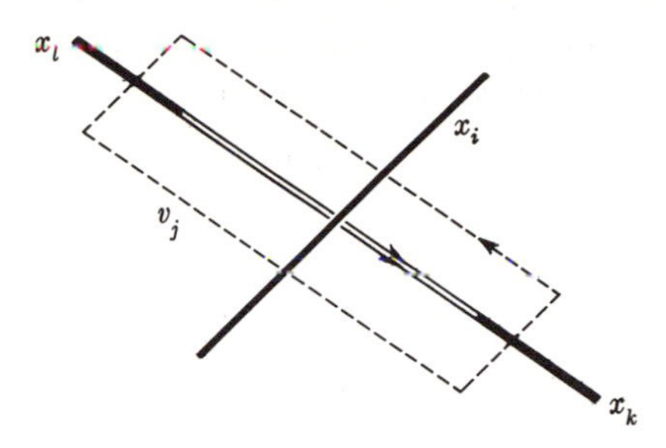

Figure 39. $v_j^{\#} = x_k^{-1}x_i^{\epsilon}x_lx_i^{-\epsilon}$, where $\epsilon = \pm 1$

4. Examples of presentations. We give below over presentations of several knots. As remarked in the preceding section, we have used (3.1) as a model instead of (1.1) because it is simpler. In addition, we shall take advantage of the fact that an arbitrary one of the relators can be dropped. The resulting presentation is frequently still needlessly complex and can be further simplified by using Tietze operations. Some of these reductions are illustrated in the following examples.

(4.1) *Trivial knot* (Figure 40).

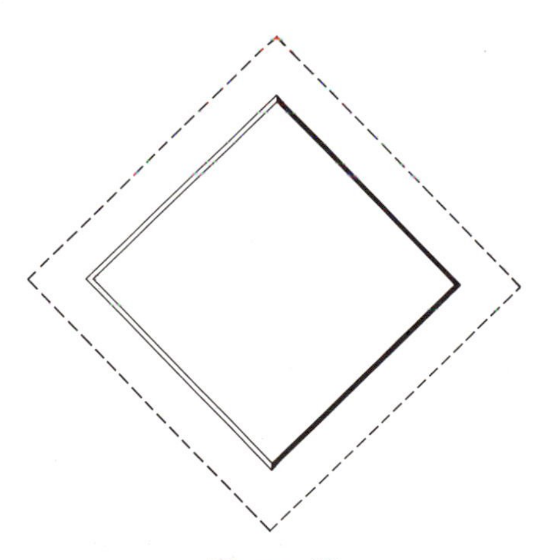

Figure 40

The single overpass is drawn with a heavy line, the underpass with a double light line, and the path v_1 with a dotted line. The presentation is

$$\pi(R^3 - K) = |\ x\ :\ |.$$

Hence, *the group of the trivial knot type is infinite cyclic.*

(4.2) *Clover-leaf knot* (Figure 41). Again, the overpasses are shown in heavy lines and the loops v_1, v_2, v_3 are drawn with dotted lines around the underpasses. We choose generators x,y,z such that

$$x = a_1^{\#}, \quad y = a_2^{\#}, \quad z = a_3^{\#}.$$

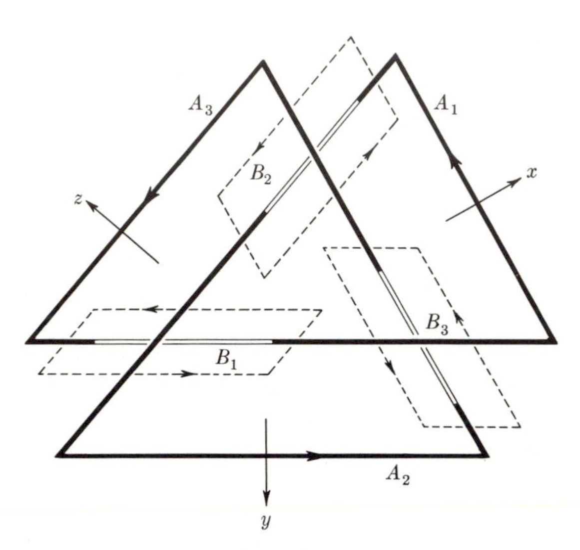

Figure 41

Clearly,

$$v_1^{\#} = x^{-1}yzy^{-1}$$
$$v_2^{\#} = y^{-1}zxz^{-1}$$
$$v_3^{\#} = z^{-1}xyx^{-1}.$$

Consequently, we obtain for the group $\pi(R^3 - K)$ of the clover-leaf knot K the presentation

$$(x,y,z \ : \ x^{-1}yzy^{-1}, y^{-1}zxz^{-1}),$$

where $v_3^{\#}$ has been dropped.

Suppose we include in the presentation all three relators obtained by reading around the underpasses. Writing relations instead of the more formal relators, we get

$$\pi(R^3 - K) = |\ x,y,z \ : \ x = yzy^{-1}, y = zxz^{-1}, z = xyx^{-1}\ |.$$

Substitution of $z = xyx^{-1}$ in the other two relations yields

$$\pi(R^3 - K) = |\ x,y \ : \ x = yxyx^{-1}y^{-1}, y = xyxy^{-1}x^{-1}\ |.$$

If the second relation is multiplied through on the right by $xyx^{-1}y^{-1}$, one obtains the first. This is empirical verification of the claim that any one of the relations obtained by reading around the underpasses of an over presentation is a consequence of the others. Finally, therefore, we obtain the following common presentation of the group $\pi(R^3 - K)$ of the clover-leaf knot:

$$(x,y \ : \ xyx = yxy).$$

(4.3) *Figure-eight knot* (Figure 42). Figure 42 shows a different projection

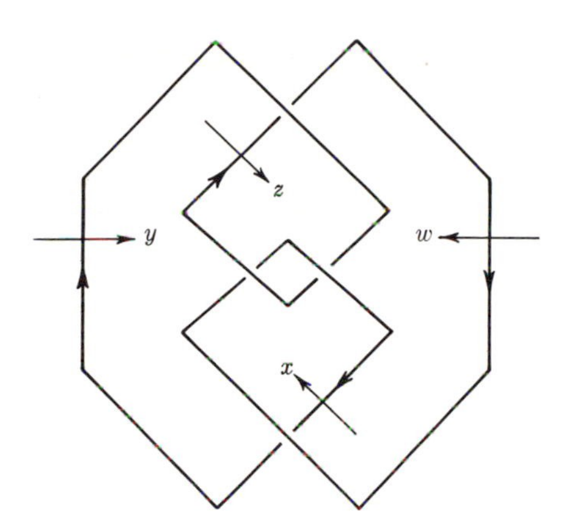

Figure 42

of the figure-eight knot from the ones given in Figures 4 and 7 of Chapter I. Using either a piece of string or a pencil and paper, one can easily show that the knot type represented below is the same. A Wirtinger presentation is

$$(x,y,z,w \ : \ x = z^{-1}wz, \ y = wxw^{-1}, \ z = x^{-1}yx).$$

Substituting $z = x^{-1}yx$ in the other two relations, we obtain

$$\pi(R^3 - K) = | \ x,y,w \ : \ x = x^{-1}y^{-1}xwx^{-1}yx, \ y = wxw^{-1} \ |.$$

The first relation now gives $w = x^{-1}yxy^{-1}x$. Substitution in the second yields

$$\pi(R^3 - K) = | \ x,y \ : \ y = x^{-1}yxy^{-1}xyx^{-1}y^{-1}x \ |.$$

Finally,

$$\pi(R^3 - K) = | \ x,y \ : \ yx^{-1}yxy^{-1} = x^{-1}yxy^{-1}x \ |.$$

(4.4) *Three-lead four bight Turk's head knot* (Figure 43).[2]

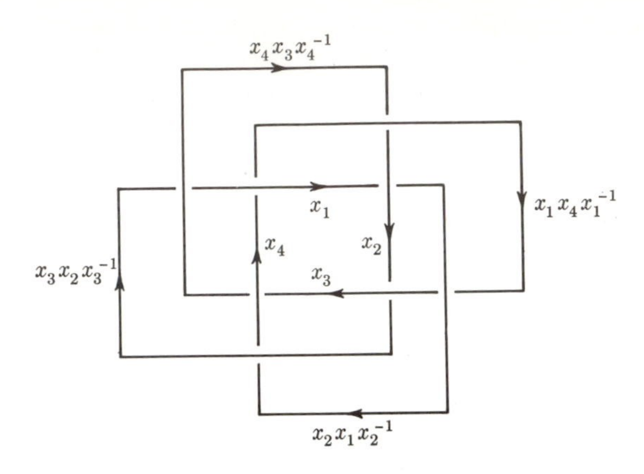

Figure 43

$$\pi(R^3 - K) = |\, x_1,x_2,x_3,x_4 \;:\; x_i = (x_{i+3}x_{i+2}^{-1}x_{i+3}^{-1})(x_{i+2}x_{i+1}x_{i+2}^{-1})(x_{i+3}x_{i+2}x_{i+3}^{-1}),$$

$$i = 1, \cdots, \hat{j}, \cdots, 4 \,|$$

$$= |\, x_1,x_2,x_3,x_4 \;:\; x_i = [x_{i+3},x_{i+2}^{-1}]x_{i+1}[x_{i+2}^{-1},x_{i+3}] \,|.$$

We recall that the element $[a,b]$ is the *commutator*, $[a,b] = aba^{-1}b^{-1}$. Notice that $[a,b]^{-1} = [b,a]$.

5. Existence of nontrivial knot types. We are now in a position to prove that different knot types exist. We shall prove that the clover-leaf knot cannot be untied, i.e., that its type is different from that of the trivial knot. We recall that if knots K and K' are of the same type, then the complementary spaces $R^3 - K$ and $R^3 - K'$ are homeomorphic and hence $\pi(R^3 - K)$ and $\pi(R^3 - K')$ are isomorphic fundamental groups. The fact that *knots of the same type have isomorphic groups* is the principle by which we shall distinguish knot types in this book. The nontriviality of the clover-leaf knot will be established by proving that its group $|\, x,y \;:\; xyx = yxy \,|$ is not infinite cyclic (cf. (4.1) and (4.2)). For this purpose, consider the symmetric group S_3 of degree 3, which is generated by the cycles (12) and (23). We observe, first of all, that S_3 is not abelian since

$$(12)(23) = (132) \neq (123) = (23)(12).$$

[2] Indices are mod 4; thus, $x_{3+2} = x_1$, etc.

The over presentation (4.2) of the clover-leaf knot consists of a homomorphism ϕ of the free group F, for which x and y constitute a free basis, onto the knot group, and the kernel of ϕ is the consequence of $xyx(yxy)^{-1}$. The homomorphism θ of F onto S_3 defined by

$$\theta(x) = (12) \quad \theta(y) = (23),$$

induces a homomorphism of the knot group onto S_3 provided $\theta(xyx) = \theta(yxy)$. But

$$\theta(xyx) = \theta(x)\,\theta(y)\,\theta(x) = (12)(23)(12) = (13),$$

$$\theta(yxy) = \theta(y)\,\theta(x)\,\theta(y) = (23)(12)(23) = (13).$$

Thus, the knot group of the clover-leaf knot can be mapped homomorphically onto a nonabelian group. It follows that the knot group itself is nonabelian, and hence it is certainly not infinite cyclic. We conclude that *the clover-leaf knot cannot be untied.*

Likewise, in order to prove that the clover-leaf is distinct from the figure-eight knot, it is sufficient to show that their groups are not isomorphic. These groups are presented in (4.2) and (4.3). Unfortunately, there is no general procedure for determining whether or not two presentations determine isomorphic groups. We do know, from the Tietze theorem, that if two groups are isomorphic, their finite presentations are related by the Tietze operations. However, attempting to show directly that the presentations given in (4.2) and (4.3) are *not* related by Tietze operations does not seem a potentially easy job.[3] What is needed are some standard procedures for deriving from a group presentation some easily calculable algebraic quantities which are the same for isomorphic groups and hence are so-called *group invariants*. That is, the group of a knot type is usually too complicated an invariant, and so we must pass to one that is simpler and easier to handle. In so doing there is a danger of throwing the baby away with the bath water. In passing to simpler invariants one invariably loses some information. What we want to do is to achieve readily distinguishable invariants which are still fine enough to distinguish the groups of at least a large number of different knots. The next two chapters are devoted to just this problem.

[3] Actually, the clover-leaf (group G) and the figure-eight knot (group G') can be distinguished by the same method we used in demonstrating the nontriviality of the clover-leaf. If $G \approx G'$, there must exist a homomorphism ξ mapping G' onto the symmetric group S_3. Using the presentation $G' = |x,y : yx^{-1}yxy^{-1} = x^{-1}yxy^{-1}x|$, and by simply exhausting the finite number of possibilities, one can check that no assignment of x and y into S_3 extends to a homomorphism of G' onto S_3. (However this does not show that the figure-eight knot cannot be untied, although this will be shown true in Chapter VIII.)

EXERCISES

1. For each of the following knots find a presentation of the group of the knot that has just two generators.

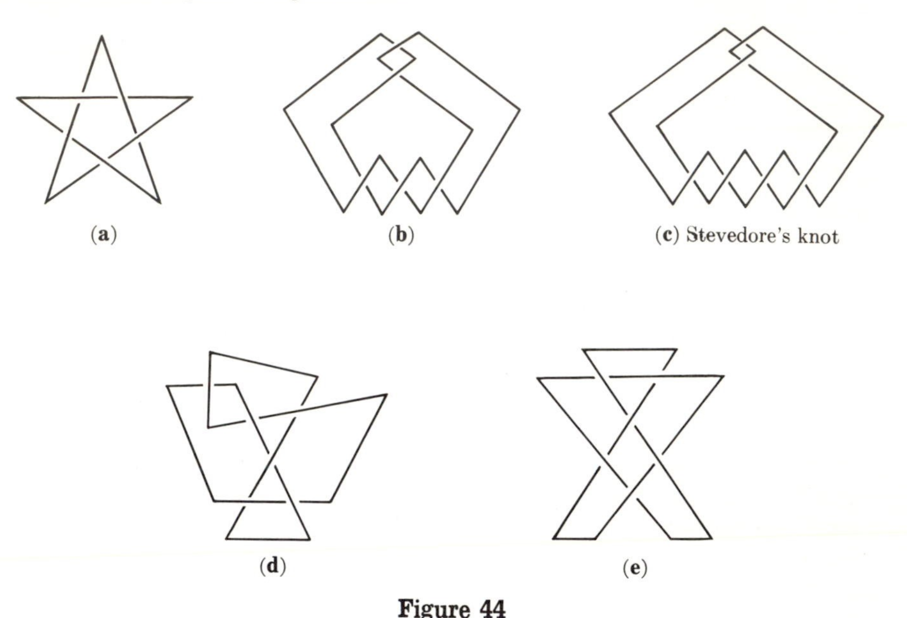

Figure 44

2. For each of the five knots considered in Exercise 1, determine the representations into the symmetric group of degree 3.

3. A torus knot $K_{p,q}$ of type p,q (where p and q are relatively prime integers) is a curve on the surface of the "unknotted" torus $(r - 2)^2 + z^2 = 1$ that cuts a meridian in p points and a longitude in q points. It is represented by the equation $r = 2 + \cos\,(q\theta/p)$, $z = \sin\,(q\theta/p)$. By dividing $R^3 - K$ into the part not interior to the torus and the part not exterior to it and applying van Kampen's theorem, prove that a presentation of the group of $K_{p,q}$ is $(a,b \;:\; a^p = b^q)$.

4. (a) Show geometrically that the clover-leaf knot is the torus knot $K_{2,3}$. (b) Draw a picture of the torus knot $K_{2,5}$.

5. Prove that the presentation of the group of the clover-leaf knot derived in (4.2) is of the same type as $(a,b \;:\; a^2 = b^3)$.

6. Let us say that a knot diagram has property l if it is possible to color the projected overpasses in three colors, assigning a color to each edge in such a way that (a) the three overpasses that meet at a crossing are either all colored the same or are all colored differently; (b) all three colors are actually used.

Show that a diagram of a knot K has property l if and only if the group of K can be mapped homomorphically *onto* the symmetric group of degree 3.

7. Show that property l is equivalent to the following: It is possible to assign an integer to each edge in such a way that the sum of the three edges that meet at any crossing is divisible by 3.

8. Show that no knot group can be represented upon the fundamental group of the Klein bottle.

9. Using the fact that the group G of the overhand knot has a presentation $(a,b \ : \ a^2 = b^3)$ find all (classes of) representations of G onto the alternating group A_5 of degree 5.

CHAPTER VII

The Free Calculus and the Elementary Ideals

Introduction. In the last chapter a method was developed for calculating a presentation of the group of any knot in regular position. Unfortunately, it does not follow, as was pointed out, that it is now a simple matter to distinguish knot groups, and thus knot types. There is no general algorithm for deciding whether or not two presentations define isomorphic groups, and even in particular examples the problem can be difficult. We are therefore concerned with deriving some powerful, yet effectively calculable, invariants of group presentation type. Such are the elementary ideals. In this chapter we shall study the necessary algebraic machinery for defining these invariants. Specialization to presentations of knot groups in Chapter VIII then leads naturally to the knot polynomials. With these invariants we can easily distinguish many knot types.

1. The group ring. With any multiplicative group G there is associated its *group ring* JG with respect to the ring J of integers. The elements of JG are the mappings $\nu\colon G \to J$ which satisfy $\nu(g) = 0$ except for at most a finite number of $g \in G$. Addition and multiplication in JG are defined respectively by

$$(\nu_1 + \nu_2)g = \nu_1 g + \nu_2 g$$

$$(\nu_1 \nu_2)g = \sum_{h \in G} (\nu_1 h)(\nu_2 h^{-1} g)$$

for any ν_1, $\nu_2 \in JG$ and $g \in G$. It is a straightforward matter to verify that JG is a ring with respect to these two operations. Multiplication by an integer n, which is defined in any ring, satisfies the equation

$$(n\nu)g = n(\nu g)$$

for any $\nu \in JG$ and $g \in G$.

There exists a mapping $G \to JG$ which assigns to each $g \in G$ the function g^* defined by

$$g^*(h) = \begin{cases} 1 \text{ if } h = g, \\ 0 \text{ otherwise.} \end{cases}$$

This mapping is one-one and product-preserving. Since the image of any product-preserving mapping of a group into a ring is a group, we conclude that the mapping $g \to g^*$ is a group isomorphism of G onto its image. Where $e \in G$ denotes the identity, the element e^* is a nonzero multiplicative identity for the ring JG.

Let ν be an arbitrary nonzero element of JG. Let $g_1, \cdots, g_k$, $k \geq 1$, be the elements of G for which $\nu(g_i) \neq 0$ and let

$$n_i = \nu(g_i), \qquad i = 1, \cdots, k.$$

Then

$$\nu = n_1 g_1^* + \cdots + n_k g_k^*.$$

That is, the image of G under the mapping $g \to g^*$ generates the additive group JG. Henceforth we identify g and g^* and write elements of JG as finite integral combinations of elements of G. It then becomes obvious, for example, that

(1.1) *JG is a commutative ring if and only if G is a commutative group.*

It is trivial to prove that if $n_1 g_1 + \cdots + n_k g_k = 0$ and $g_1, \cdots, g_k$ are distinct, then $n_1 = \cdots = n_k = 0$. It follows that a nonzero element of JG can be written as a finite sum of distinct group elements with nonzero coefficients in one and only one way. This fact implies that, as an additive group, JG satisfies the analogue of the characteristic property of free groups that was discussed in Chapter III:

(1.2) *An arbitrary mapping ϕ of G into an additive abelian group A has a unique extension to an additive homomorphism ϕ: $JG \to A$. Moreover, if A is a ring and ϕ preserves products on G, the extension is a ring homomorphism.*

Proof. Set $\phi 0 = 0$. Every nonzero element of JG has a unique expression $n_1 g_1 + \cdots + n_k g_k$ where $n_i \neq 0$, $i = 1, \cdots, k$, and $g_1, \cdots, g_k$ are distinct. To obtain the extension, we define

$$\phi(n_1 g_1 + \cdots + n_k g_k) = n_1 \phi g_1 + \cdots + n_k \phi g_k.$$

Next, we observe that if this equation holds under the imposed conditions, it also holds for arbitrary integers $n_1, \cdots, n_k$ and group elements $g_1, \cdots, g_k$. But then, ϕ is trivially addition-preserving. Since any extension of ϕ to an additive homomorphism $JG \to A$ must satisfy this equation, the uniqueness of the extension is assured. Finally, if A is a ring and ϕ is product-preserving in G,

$$\begin{aligned} \phi(\sum_i n_i g_i \sum_j n_j' g_j') &= \phi(\sum_{ij} n_i n_j' g_i g_j') \\ &= \sum_{ij} n_i n_j' \phi g_i \phi g_j' = \sum_i n_i \phi g_i \sum_j n_j' \phi g_j' \\ &= (\phi \sum_i n_i g_i)(\phi \sum_j n_j' g_j'), \end{aligned}$$

and the proof is complete.

As an additive group, therefore, JG is free abelian and the subset G is a basis. (Cf. Section 5, Chapter IV.) Another corollary of (1.2) is

(1.3) *Every group homomorphism ϕ: $G \to G'$ has a unique extension to a ring homomorphism ϕ: $JG \to JG'$.*

Notice that if any one of the restrictions; (i) ϕ is the identity mapping; (ii) ϕ is onto; (iii) ϕ is one-one, is satisfied by the group homomorphism ϕ: $G \to G'$, then it also holds for the ring homomorphism ϕ: $JG \to JG'$.

Two homomorphisms defined on the group ring of every group come up sufficiently often to warrant special mention and notation. The first of these is the abelianizer, which was introduced in Section 4 of Chapter IV as the canonical homomorphism of a group onto its commutator quotient group. By the *abelianizer* we shall now mean either the group homomorphism $\mathfrak{a}$: $G \to G/[G,G]$, as before, or its unique extension on the group rings. The second homomorphism is the *trivializer*. For any group G consider the mapping $\mathfrak{t}$: $G \to J$ defined by $\mathfrak{t}(g) = 1$, for all $g \in G$. The trivializer is the unique extension of $\mathfrak{t}$ to the ring homomorphism $\mathfrak{t}$: $JG \to J$. Clearly,

$$\mathfrak{t}(\sum_i n_i g_i) = \sum_i n_i.$$

We conclude this section with the observation that the mapping $J \to JG$ defined by $n \to ne$, where $e \in G$ is the identity, is a ring isomorphism. Hence, both G and J are considered as subsets of the group ring JG.

2. The free calculus. By a *derivative* in a group ring JG will be meant any mapping D: $JG \to JG$ which satisfies

$$(2.1) \qquad D(\nu_1 + \nu_2) = D\nu_1 + D\nu_2,$$

$$(2.2) \qquad D(\nu_1\nu_2) = (D\nu_1)\mathfrak{t}(\nu_2) + \nu_1 D\nu_2,$$

where $\mathfrak{t}$ is the trivializer and $\nu_1, \nu_2 \in JG$. For elements of G, (2.2) takes the simpler form

$$(2.3) \qquad D(g_1g_2) = Dg_1 + g_1Dg_2, \qquad g_1, g_2 \in G.$$

In fact, in view of (1.2), a derivative may be defined as the unique linear extension to JG of any mapping D of G into JG which satisfies (2.3). Obviously, the constant mapping of JG onto the element 0 is a derivative. The question of whether or not nontrivial derivatives always exist in an arbitrary group ring is settled by the observation that the mapping $g \to g - 1$, for any $g \in G$, determines a derivative. Moreover, if D and D' are two derivatives in JG and $\nu_0 \in JG$ is an arbitrary element, it is easy to check that the mappings $D + D'$ and $D \circ \nu_0$ defined respectively by

$$(2.4) \qquad \left.\begin{aligned} (D + D')\nu &= D\nu + D'\nu \\ (D \circ \nu_0)\nu &= (D\nu)\nu_0 \end{aligned}\right\} \nu \in JG$$

are also derivatives in JG.

It is worthwhile to note some of the consequences of the axioms for a derivative:

(2.5) $$D(\Sigma n_i g_i) = \Sigma n_i Dg_i.$$

(2.6) $$Dn = 0, \textit{ for any integer } n.$$

(2.7) $$Dg^{-1} = -g^{-1}Dg, \textit{ for any } g \in G.$$

The first follows from the fact that D is an additive homomorphism. Since $D1 = D(1 \cdot 1) = D1 + D1$, we conclude that $D1 = 0$ and, more generally, that (2.6) holds. Lastly, the equations $0 = D(g^{-1} g) = Dg^{-1} + g^{-1}Dg$ establish (2.7). It is useful to define, for any $g \in G$ and integer n, the group ring element

$$\frac{g^n - 1}{g - 1} = \begin{cases} 0, & \text{if } n = 0, \\ \sum_{i=0}^{n-1} g^i, & \text{if } n > 0, \\ -\sum_{i=n}^{-1} g^i, & \text{if } n < 0. \end{cases}$$

It follows that

(2.8) $$Dg^n = \frac{g^n - 1}{g - 1} Dg.$$

The proof is by induction on the absolute value of n. For $n = 0, +1, -1$, the assertion reduces respectively to $D1 = 0$, $Dg = Dg$, and $Dg^{-1} = -g^{-1}Dg$. Next assume $| n | > 1$. If n is positive, (2.3) and the hypothesis of induction yield

$$\begin{aligned} Dg^{n+1} &= Dg^n + g^n Dg \\ &= \sum_{i=0}^{n-1} g^i Dg + g^n Dg \\ &= \frac{g^{n+1} - 1}{g - 1} Dg. \end{aligned}$$

Similarly, for negative n,

$$\begin{aligned} Dg^{n-1} &= Dg^n + g^n Dg^{-1} \\ &= -\sum_{i=n}^{-1} g^i Dg - g^{n-1} Dg \\ &= \frac{g^{n-1} - 1}{g - 1} Dg, \end{aligned}$$

and the proof is complete.

Another consequence of the axioms is the fact that any derivative is uniquely determined by its values on any generating subset of G.

Although we have introduced the notion of a derivative in an arbitrary group ring, we are here really interested in, and henceforth shall confine our attention to, derivatives in the group ring of a free group (hence the title of this section). Suppose F is a free group with a free basis of elements $x_1, x_2, \cdots$. An arbitrary element of the group ring JF is a finite sum of finite products of powers of these x's, and it is natural to think of such elements as polynomials in the variables $x_1, x_2, \cdots$. Of course they are not true polynomials since negative powers may occur and, what is more important, the variables do not commute. Nevertheless, an element of JF will be called a *free polynomial* and will be denoted generically as $f(x) = \Sigma n_i u_i$, $u_i \in F$. The action of the trivializer $\mathfrak{t}$ on a free polynomial is indicated by writing $\mathfrak{t}f(x) = f(1)$.

(2.9) *To each free generator x_j there corresponds a unique derivative*

$$D_j = \partial/\partial x_j$$

in JF, called the derivative with respect to x_j, which has the property

$$\frac{\partial x_i}{\partial x_j} = \delta_{ij} \text{ (Kronecker delta).}$$

Proof. Since the values of the derivative $\partial/\partial x_j$ are specified on a generating subset of F, uniqueness is automatic. In order to prove existence, we consider an arbitrary set A of elements $a_1, a_2, \cdots$ in one-one correspondence with $x_1, x_2, \cdots$ under the assignment $\theta a_i = x_i$. From the results of Chapter III we know that θ extends to a product-preserving mapping of the semi-group $W(A)$ of words onto the group F under which equivalent words in $W(A)$ map onto the same group element in F. We propose to define, for each x_j, a mapping Λ_j: $W(A) \to JF$ which will induce the derivative $\partial/\partial x_j$. The definition is by induction on the number of syllables in a word and is given by

$$(*)\qquad \begin{cases} \Lambda_j 1 = 0 \text{ (where 1 denotes the empty word)}, \\ \Lambda_j a_i^n = \dfrac{x_i^n - 1}{x_i - 1}\,\delta_{ij}, \\ \Lambda_j(a_i^n a) = \Lambda_j a_i^n + x_i^n \Lambda_j a, \qquad a \in W(A). \end{cases}$$

The reason for retreating to the semi-group $W(A)$ for the basic definition should be clear. With the absence of any cancellation law in $W(A)$, the function Λ_j is unambiguously defined by (*). We contend that

$$(**)\qquad \Lambda_j(ab) = \Lambda_j a + \theta a \cdot \Lambda_j b, \qquad a,b \in W(A).$$

The proof is by induction on the number of syllables in ab. If a is the empty word, the result holds trivially; so we assume that a contains at least one syllable. Hence, $a = a_i^n c$ and by (*)

$$\Lambda_j(ab) = \Lambda_j(a_i^n cb) = \Lambda_j a_i^n + x_i^n \Lambda_j(cb).$$

By the hypothesis of induction, therefore,

$$\Lambda_j(ab) = \Lambda_j a_i^n + x_i^n \Lambda_j c + x_i^n \theta c \cdot \Lambda_j b = \Lambda_j a + \theta a \cdot \Lambda_j b.$$

We next assert that if two words are equivalent, then their images under Λ_j are equal. The proof amounts to verifying

$$\Lambda_j(aa_i{}^0b) = \Lambda_j(ab),$$
$$\Lambda_j(aa_i^{m+n}b) = \Lambda_j(aa_i{}^m a_i{}^n b).$$

With (**) available, these identities follow easily. For the first, we have

$$\begin{aligned}\Lambda_j(aa_i{}^0b) &= \Lambda_j(aa_i{}^0) + \theta(aa_i{}^0)\cdot\Lambda_j b\\ &= \Lambda_j a + \theta a\cdot\Lambda_j b\\ &= \Lambda_j(ab).\end{aligned}$$

For the second, we note first that

$$\frac{x_i^{m+n}-1}{x_i-1} = \frac{x_i{}^m-1}{x_i-1} + x_i{}^m\,\frac{x_i{}^n-1}{x_i-1}\,.$$

Thus, $\Lambda_j a_i^{m+n} = \Lambda_j(a_i{}^m a_i{}^n)$; hence,

$$\begin{aligned}\Lambda_j(aa_i^{m+n}b) &= \Lambda_j(aa_i^{m+n}) + \theta(aa_i^{m+n})\cdot\Lambda_j b\\ &= \Lambda_j a + \theta a\cdot\Lambda_j(a_i{}^m a_i{}^n) + \theta(aa_i{}^m a_i{}^n)\cdot\Lambda_j b\\ &= \Lambda_j(aa_i{}^m a_i{}^n b),\end{aligned}$$

and the assertion is proved. The mapping $\partial/\partial x_j\colon F \to JF$ is now defined by

$$(***)\qquad \frac{\partial}{\partial x_j}\,\theta a = \Lambda_j a, \qquad \text{for any} \quad a \in W(A).$$

The function $\partial/\partial x_j$ is well-defined because θ is onto and because if $\theta a = \theta b$, then a and b are equivalent words and so $\Lambda_j a = \Lambda_j b$. (Notice that the proof that Λ_j depends only on equivalence classes of words depends on the fact that the same is true of θ.) It follows immediately from (***) that $\partial x_i/\partial x_j = \delta_{ij}$. To check that $\partial/\partial x_j$ determines a derivative in JF, we have only to verify Axiom (2.3): For any $u = \theta a$ and $v = \theta b$ in F,

$$\frac{\partial}{\partial x_j}(uv) = \frac{\partial}{\partial x_j}\theta(ab) = \Lambda_j(ab) = \Lambda_j a + u\Lambda_j b = \frac{\partial}{\partial x_j}\,u + u\,\frac{\partial}{\partial x_j}\,v,$$

and the proof of (2.9) is complete.

The preceding theorem is a remarkable result in that it reveals the entire structure of the set of derivatives in a free group ring. This assertion is formulated explicitly in the following important corollary:

(2.10) *For any free polynomials* $h_1(x), h_2(x), \cdots$, *there is one and only one derivative* D *in* JF *such that* $Dx_j = h_j(x)$, $j = 1, 2, \cdots$. *Moreover, for any* $f(x) \in JF$,

$$Df(x) = \sum_j \frac{\partial f}{\partial x_j}\,h_j(x).$$

Proof. Again, uniqueness is automatic. It should next be checked that the above sum is a finite one. We leave it to the reader to prove that *if the generator x_j does not occur in the free polynomial $f(x)$, then* $\dfrac{\partial f(x)}{\partial x_j} = 0$. (cf. Exercise 5). It then follows that the mapping $f(x) \to \sum_j \dfrac{\partial f}{\partial x_j} h_j(x)$ is a derivative in JF (cf. (2.4)) with the desired property, and the proof is complete.

We have already remarked that the mapping $f(x) \to f(x) - f(1)$ is a derivative in JF. As a corollary of (2.10), we thus obtain the *fundamental formula*

$$f(x) - f(1) = \sum_j \frac{\partial f}{\partial x_j}(x_j - 1). \tag{2.11}$$

It follows that a free polynomial is determined by its derivatives and its value at 1. The analogy with the familiar law of the mean is obvious and is also capable of further interesting generalization.

3. The Alexander matrix. The free calculus is the principal mathematical tool in our construction of useful invariants of group presentation types. Consider a group presentation $(\mathbf{x} : \mathbf{r})$. The set $\mathbf{x} = (x_1, x_2, \cdots)$ is a free basis of the free group F, and the group of the presentation is the factor group

$$| \mathbf{x} : \mathbf{r} | = F/R \xleftarrow{\gamma} F$$

where R is the consequence of $\mathbf{r} = (r_1, r_2, \cdots)$ and γ is the canonical homomorphism. Both γ and the abelianizer $\mathfrak{a}$ possess unique extensions to homomorphisms of their respective group rings. Denoting the abelianized group of $| \mathbf{x} : \mathbf{r} |$ by H, we thus have the composition

$$JF \xrightarrow{\frac{\partial}{\partial x_j}} JF \xrightarrow{\gamma} J\,| \mathbf{x} : \mathbf{r} | \xrightarrow{\mathfrak{a}} JH.$$

The *Alexander matrix* of $(\mathbf{x} : \mathbf{r})$ is the matrix $\| a_{ij} \|$ defined by the formula

$$a_{ij} = \mathfrak{a}\gamma\left(\frac{\partial r_i}{\partial x_j}\right).$$

The effect of $\mathfrak{a}\gamma$ on $\partial r_i/\partial x_j$ is an immense simplification. The homomorphism γ carries elements of JF into $J| \mathbf{x} : \mathbf{r} |$ where every consequence of $\mathbf{r}$ equals 1. Even more important is the fact that $\mathfrak{a}$ then takes everything into a commutative ring. In a commutative ring one can define determinants, and furthermore, for knot groups, JH is particularly simple.

It should be remarked that the definition of the Alexander matrix assumes an ordering of the generators and relators whereas the original definition of a group presentation in Chapter IV did not. This is true but unimportant. We shall see in the next section that two matrices which differ only by a permutation of the rows or of the columns are considered equivalent anyway. The additional property of order could, of course, have been ascribed to group presentations in the first place. It is, however, (for us) an inessential feature and the definition seems simpler without it.

4. The elementary ideals. Let R be an arbitrary commutative ring with a nonzero multiplicative identity 1, and consider an m (row) $\times$ n (column) matrix A with entries in R. For any non-negative integer k, we define the kth *elementary ideal* $E_k(A)$ of A as follows:

If $0 < n - k \leq m$, *then* $E_k(A)$ *is the ideal generated by the determinants of all* $(n - k) \times (n - k)$ *submatrices of* A.

If $n - k > m$, *then* $E_k(A) = 0$.

If $n - k \leq 0$, *then* $E_k(A) = R$.

Since the determinant of any matrix can be expanded as a combination of the cofactors of the elements of any row or column, we have immediately

(4.1) *The elementary ideals of* A *form an ascending chain*

$$E_0(A) \subset E_1(A) \subset \cdots \subset E_n(A) = E_{n+1}(A) = \cdots = R.$$

If A and A' are two matrices with entries in R, we define A to be *equivalent* to A', denoted $A \sim A'$, if there exists a finite sequence of matrices

$$A = A_1, \cdots, A_n = A',$$

such that A_{i+1} is obtained from A_i, or vice-versa, by one of the following operations:

(i) *Permuting rows or permuting columns.*

(ii) *Adjoining a row of zeros,* $A \to \left\| \begin{matrix} A \\ 0 \end{matrix} \right\|$.

(iii) *Adding to a row a linear combination of other rows.*

(iv) *Adding to a column a linear combination of other columns.*

(v) *Adjoining a new row and column such that the entry in the intersection of the new row and column is* 1, *and the remaining entries in the new row and column are all* 0, $A \to \left\| \begin{matrix} A & 0 \\ 0 & 1 \end{matrix} \right\|$.

It is not hard to show that a matrix A is equivalent to the matrix obtained from A by adjoining a new row and column such that the entry in the intersection of the new row and column is 1, the remaining entries in the new column are all 0, and the remaining entries in the new row are arbitrary. Hence, (v) may be replaced by the stronger

(v') $\quad A \longrightarrow \left\| \begin{matrix} A & 0 \\ a & 1 \end{matrix} \right\|$.

The proof of this assertion is obtained in one application of (*v*) and n applications of (*iv*), where n is the number of columns in A. Thus,

$$A \xrightarrow{(v)} \left\| \begin{matrix} A & 0 \\ 0 & 1 \end{matrix} \right\| \xrightarrow{(iv)} \left\| \begin{matrix} A & 0 \\ a & 1 \end{matrix} \right\|.$$

The present definition of matrix equivalence differs from the one we usually encounter in linear algebra most notably in respect to (*ii*) and (*v*). The familiar operations of multiplication of a row and of a column by a unit e of the ring R, however, still preserve equivalence. They may be accomplished as follows:

$$(vi) \quad \left\| \begin{matrix} A \\ a \end{matrix} \right\| \xrightarrow{(ii)} \left\| \begin{matrix} A \\ a \\ 0 \end{matrix} \right\| \xrightarrow{(iii)} \left\| \begin{matrix} A \\ a \\ ea \end{matrix} \right\| \xrightarrow{(i)} \left\| \begin{matrix} A \\ ea \\ a \end{matrix} \right\| \xrightarrow{(iii)} \left\| \begin{matrix} A \\ ea \\ a - e^{-1}\, ea \end{matrix} \right\| \xleftarrow{(ii)} \left\| \begin{matrix} A \\ ea \end{matrix} \right\|.$$

$$(vii) \quad \left\| \begin{matrix} A & a \end{matrix} \right\| \xrightarrow{(v')} \left\| \begin{matrix} A & a & 0 \\ 0 & -e^{-1} & 1 \end{matrix} \right\| \xrightarrow{(iv)} \left\| \begin{matrix} A & a & ea \\ 0 & -e^{-1} & 0 \end{matrix} \right\|$$

$$\xrightarrow{(i)} \left\| \begin{matrix} A & ea & a \\ 0 & 0 & -e^{-1} \end{matrix} \right\| \xrightarrow{(iv)} \left\| \begin{matrix} A & ea & 0 \\ 0 & 0 & -e^{-1} \end{matrix} \right\|$$

$$\xrightarrow{(vi)} \left\| \begin{matrix} A & ea & 0 \\ 0 & 0 & 1 \end{matrix} \right\| \xleftarrow{(v)} \left\| \begin{matrix} A & ea \end{matrix} \right\|.$$

It should also be observed that matrix equivalence, as we have defined it, is trivially reflexive, symmetric, and transitive. That is, it is a true equivalence relation.

(4.2) *Equivalent matrices define the same chain of elementary ideals.*

Proof. The proof depends on the well-known elementary facts of determinant manipulation. Incidentally, these are purely combinatorial in nature and hold in any commutative ring. We must prove that $E_k(A) = E_k(A')$ where A' is obtained from A by any one of the above operations (*i*), $\cdots$, (*v*). For (*i*), (*iii*), and (*iv*), the result is immediate. For example, consider (*iii*): Any generator of $E_k(A')$ is either already a generator of $E_k(A)$ or, by expansion by the minors of a row, is a linear combination of generators of $E_k(A)$. The same expansion shows that the converse proposition is valid.

Next, consider operation (*ii*): Since $n = n'$ and $m' = m + 1$, we see that if $0 < n - k \leq m$, then $0 < n' - k \leq m'$, and in this range we have $E_k(A) = E_k(A')$. The only other possibility that is not definitionally immediate is $n - k = m'$. In this case, $E_k(A) = 0$ follows trivially and $E_k(A')$ is generated by the determinants of the $m' \times m'$ submatrices of A'. Since the last row of each of these submatrices contains all zeros, $E_k(A') = 0$ holds as well.

Finally, we must check operation (*v*). Here $n' = n + 1$ and $m' = m + 1$. If $n - k > m$, then $n' - k = (n + 1) - k > m + 1 = m'$ and the identities $E_k(A) = E_k(A') = 0$ are immediate. If $n - k \leq 0$, then

$n' - k \leq 1$. In this case, $E_k(A) = R$ follows trivially, and the same is true for A' except when $n' - k = 1$. But $E_{n'-1}(A')$ is the ideal generated by the elements of A'. Since one of these elements is 1, we conclude $E_{n'-1}(A') = R$. The remaining range is $0 < n - k \leq m$. Any $(n - k) \times (n - k)$ submatrix of A can be enlarged to an $(n' - k) \times (n' - k)$ submatrix of A' by adjoining appropriate elements of the new row and column of A' including thereby the element 1 in the intersection. Expansion by the minors of the new column shows that the determinants of these two submatrices are equal, except possibly for sign. Thus, $E_k(A) \subset E_k(A')$. Conversely, consider any $(n' - k) \times (n' - k)$ submatrix of A' whose rows may or may not include the new row of A' (note $n' - k \geq 2$). If they do, expansion by minors of the row shows that its determinant is a generator of $E_k(A)$ and, consequently, belongs to $E_k(A)$. If its rows do not include the new row of A', its determinant belongs to $E_{k-1}(A)$, which by (4.1) is contained in $E_k(A)$. We conclude that, under operation (v), $E_k(A) \supset E_k(A')$, and the proof is complete.

Consider an arbitrary ring homomorphism $\phi\colon R \to R'$ where R and R' are any two commutative rings containing multiplicative identities. If $A = \| a_{ij} \|$ is a matrix with entries in R, we define the image matrix

$$\phi A = \| \phi(a_{ij}) \|.$$

A useful result is

(4.3) *If ϕ is onto, then $\phi E_k(A) = E_k(\phi A)$.*

Proof. Of course, $\phi(0) = 0$ automatically; but the equation $\phi R = R'$, which is needed if $n - k \leq 0$, is just the hypothesis that ϕ is onto. Obviously, the image of the set of determinants of all $(n - k) \times (n - k)$ submatrices of A equals the set of all determinants of all $(n - k) \times (n - k)$ submatrices of ϕA. Thus, we have only to ask whether the image of an ideal generated by a set of elements $a_1, \cdots, a_r$ in R is the ideal generated by the images $\phi(a_1), \cdots, \phi(a_r)$ in R'. The answer is easily seen to be yes; but, again, only if ϕ is onto.

For any finite group presentation $(\mathbf{x} : \mathbf{r})$ and non-negative integer k, we define the kth *elementary ideal of* $(\mathbf{x} : \mathbf{r})$ to be the kth elementary ideal of the Alexander matrix of $(\mathbf{x} : \mathbf{r})$. By virtue of (4.2), of course, the elementary ideals of a presentation may be calculated from any matrix equivalent to the Alexander matrix. In any specific example one naturally uses the simplest matrix one can find.

The elementary ideals, defined for any finite group presentation, represent a generalization of the knot polynomials which we shall define in the next chapter for presentations of knot groups. There are several reasons for introducing the ideals before the polynomials. First of all, whereas the ideals are defined for arbitrary finitely presented groups, the polynomials exist and are unique only for a more restricted class of groups satisfying certain algebraic conditions. In the next chapter we shall discuss these conditions and show that any tame knot group satisfies them. Furthermore, it is, if anything, easier to prove the invariance of the ideals than of the polynomials. Since the latter can be characterized in terms of the former, we can kill two birds with one stone. Finally, even where the polynomials do exist, the ideals contain more information. We shall exhibit two knots in Chapter VIII which are not distinguishable by their polynomials, but which do have different elementary ideals.

The immediate problem is to prove that the elementary ideals of a finite presentation are invariants of the type of the presentation. The proof is based on the Tietze theorem, which reduces the problem to checking the invariance of the ideals under Tietze operations **I** and **II**. The essential part of the proof, therefore, is simply examining the effect on the Alexander matrices of each of these operations. For a clear understanding of the formulation of the invariance theorem which appears below, the reader may wish to review the basic definitions and results on presentation mappings and the Tietze theorem in Chapter IV.

If $f\colon (\mathbf{x} : \mathbf{r}) \to (\mathbf{y} : \mathbf{s})$ is an arbitrary presentation mapping, there is induced a homomorphism $f_*\colon |\mathbf{x} : \mathbf{r}| \to |\mathbf{y} : \mathbf{s}|$ on the groups of the presentations. This mapping in turn induces a homomorphism f_{**} of the abelianized group of $|\mathbf{x} : \mathbf{r}|$ into that of $|\mathbf{y} : \mathbf{s}|$ (cf. (4.1), Chapter IV). If $(\mathbf{x} : \mathbf{r})$ and $(\mathbf{y} : \mathbf{s})$ are known to be of the same type, there exists a presentation equivalence $(\mathbf{x} : \mathbf{r}) \xrightarrow{f} (\mathbf{y} : \mathbf{s}) \xrightarrow{g} (\mathbf{x} : \mathbf{r})$ and

$$\text{identity} = (fg)_* = f_*g_*,$$

$$\text{identity} = (f_*g_*)_* = f_{**}g_{**}.$$

Similarly, $g_{**}f_{**}$ is the identity. Thus,

(4.4) *If the pair f, g is a presentation equivalence, then each of f_{**} and g_{**} is an isomorphism onto and the inverse of the other.*

Notice that the same conclusion holds for the extensions of f_{**} and g_{**} to their respective ring homomorphisms on the appropriate group rings.

We recall that if the pair of mappings f,g is a presentation equivalence, then each of f and g may be called a presentation equivalence separately. To speak of one alone, however, always implies the existence of a mate. The statement that the elementary ideals of a finite group presentation are invariants of the presentation type is the following theorem:

(4.5) THE INVARIANCE OF THE ELEMENTARY IDEALS. *If $(\mathbf{x} : \mathbf{r})$ and $(\mathbf{y} : \mathbf{s})$ are finite group presentations and*

$$f\colon (\mathbf{x} : \mathbf{r}) \to (\mathbf{y} : \mathbf{s})$$

is a presentation equivalence, then the kth elementary ideal of $(\mathbf{x} : \mathbf{r})$ is mapped by f_{**} onto the kth elementary ideal of $(\mathbf{y} : \mathbf{s})$.

Proof. As a result of the Tietze theorem, the proof immediately reduces to checking only the Tietze equivalences **I**, **I′**, **II**, and **II′**. Also observe that, in view of (4.4), if (4.5) holds for one member of a pair of presentation equivalences, it holds automatically for the other. So we need check only Tietze **I** and **II**.

Tietze **I**. This is the presentation mapping

$$(\mathbf{x} : \mathbf{r}) \xrightarrow{\mathbf{I}} (\mathbf{x} : \mathbf{r} \cup s),$$

where $\mathbf{x} = (x_1, \cdots, x_n)$, $\mathbf{r} = (r_1, \cdots, r_m)$, the element s is a consequence of $\mathbf{r}$, and $\mathbf{I}$: $F(\mathbf{x}) \to F(\mathbf{x})$ is the identity mapping. It follows that $\mathbf{I}_*$ and $\mathbf{I}_{**}$ are also identities. Hence, the argument is completed by simply showing that $(\mathbf{x} : \mathbf{r})$ and $(\mathbf{x} : \mathbf{r} \cup s)$ have equivalent Alexander matrices. Since s is a consequence of $\mathbf{r}$, we have

$$s = \prod_{k=1}^{p} u_k r_{i_k}^{\alpha_k} u_k^{-1},$$

$$\frac{\partial s}{\partial x_j} = \frac{\partial}{\partial x_j}(u_1 r_{i_1}^{\alpha_1} u_1^{-1}) + u_1 r_{i_1}^{\alpha_1} u_1^{-1} \cdot \frac{\partial}{\partial x_j}(u_2 r_{i_2}^{\alpha_2} u_2^{-1}) + \cdots + \prod_{k=1}^{p-1}(u_k r_{i_k}^{\alpha_k} u_k^{-1}) \cdot \frac{\partial}{\partial x_j}(u_p r_{i_p}^{\alpha_p} u_p^{-1}).$$

But since

$$\gamma(r_i) = 1,$$

$$\gamma\left(\frac{\partial s}{\partial x_j}\right) = \sum_{k=1}^{p} \gamma\left(\frac{\partial}{\partial x_j}(u_k r_{i_k}^{\alpha_k} u_k^{-1})\right).$$

However,

$$\frac{\partial}{\partial x_j}(u_k r_{i_k}^{\alpha_k} u_k^{-1}) = \frac{\partial u_k}{\partial x_j} + u_k \frac{r_{i_k}^{\alpha_k} - 1}{r_{i_k} - 1} \frac{\partial r_{i_k}}{\partial x_j} - u_k r_{i_k}^{\alpha_k} u_k^{-1} \frac{\partial u_k}{\partial x_j},$$

and

$$\gamma\left(\frac{r_{i_k}^{\alpha_k} - 1}{r_{i_k} - 1}\right) = \alpha_k.$$

Hence,

$$\gamma\left(\frac{\partial}{\partial x_j}(u_k r_{i_k}^{\alpha_k} u_k^{-1})\right) = \alpha_k \gamma(u_k) \gamma\left(\frac{\partial r_{i_k}}{\partial x_j}\right).$$

Setting $\alpha_k \mathfrak{a}\gamma(u_k) = c_k$, we obtain finally

$$\mathfrak{a}\gamma\left(\frac{\partial s}{\partial x_j}\right) = \sum_{k=1}^{p} c_k \mathfrak{a}\gamma\left(\frac{\partial r_{i_k}}{\partial x_j}\right).$$

Thus, the Alexander matrix of $(\mathbf{x} : \mathbf{r} \cup s)$ is like that of $(\mathbf{x} : \mathbf{r})$ except for having one additional row which is a linear combination of the other rows. So the two matrices are equivalent and the first part of the proof is complete.

Tietze **II**. This is the presentation mapping

$$(\mathbf{x} : \mathbf{r}) \xrightarrow{\text{II}} (\mathbf{x} \cup y : \mathbf{r} \cup y\xi^{-1}),$$

where $\mathbf{x} = (x_1, \cdots, x_n)$, $\mathbf{r} = (r_1, \cdots, r_m) \subset F(\mathbf{x})$, $\xi \in F(\mathbf{x})$, y is a member of the underlying set of generators not contained in $\mathbf{x}$, and $\mathbf{II}$: $F(\mathbf{x}) \to F(\mathbf{x},y)$ is the inclusion mapping. Setting

$$G = | \mathbf{x} : \mathbf{r} | \text{, and } G' = | \mathbf{x} \cup y : \mathbf{r} \cup y\xi^{-1} | ,$$

and H and H' equal to the abelianized groups of G and G' respectively, we have the following array of homomorphisms:

$$\begin{array}{ccc} JF(\mathbf{x}) & \xrightarrow{\text{II}} & JF(\mathbf{x} \cup y) \\ \gamma \downarrow & & \downarrow \gamma' \\ JG & \xrightarrow{\text{II}_*} & JG' \\ \mathfrak{a} \downarrow & & \downarrow \mathfrak{a}' \\ JH & \xrightarrow{\text{II}_{**}} & JH' \end{array} \qquad \begin{aligned} \gamma'\mathbf{II} &= \mathbf{II}_*\gamma, \\ \mathfrak{a}'\mathbf{II}_* &= \mathbf{II}_{**}\mathfrak{a}. \end{aligned}$$

We denote the Alexander matrices of $(\mathbf{x} : \mathbf{r})$ and $(\mathbf{x} \cup y : \mathbf{r} \cup y\xi^{-1})$ by $A = \| a_{ij} \|$ and $A' = \| a_{ij}' \|$, respectively. Then

$$a_{ij} = \mathfrak{a}\gamma\left(\frac{\partial r_i}{\partial x_j}\right), \qquad \begin{array}{l} i = 1, \cdots, m, \\ j = 1, \cdots, n, \end{array}$$

and

$$\begin{aligned} \mathbf{II}_{**}a_{ij} = \mathbf{II}_{**}\mathfrak{a}\gamma\left(\frac{\partial r_i}{\partial x_j}\right) &= \mathfrak{a}'\gamma'\mathbf{II}\left(\frac{\partial r_i}{\partial x_j}\right) \\ &= \mathfrak{a}'\gamma'\left(\frac{\partial r_i}{\partial x_j}\right) = a_{ij}'. \end{aligned}$$

Clearly,

$$\frac{\partial r_i}{\partial y} = 0, \qquad \text{and} \qquad \frac{\partial}{\partial y}(y\xi^{-1}) = 1.$$

So, if we denote the row of elements

$$\mathfrak{a}'\gamma'\left(\frac{\partial}{\partial x_j} y\xi^{-1}\right), j = 1, \cdots, n, \text{ by } a', \text{ we have}$$

$$A' = \left\| \begin{array}{cc} \mathbf{II}_{**}A & 0 \\ a' & 1 \end{array} \right\| .$$

Hence, by matrix operation (v') of page 101,

$$A' \sim \mathbf{II}_{**}A.$$

It follows from (4.2) and (4.3) that

$$E_k(A') = E_k(\mathbf{II}_{**}A) = \mathbf{II}_{**}E_k(A),$$

and the proof of the invariance theorem is complete.

A characteristic of the techniques of Chapter IV and of our approach to the Alexander matrix and the elementary ideals in this chapter has been the construction of a theory of group presentations independent of the particular groups from which the presentations may have come (e.g., in knot theory from the fundamental groups). Thus, we have defined both the Alexander matrix and the elementary ideals of a *presentation* rather than of a *presentation of a group G*. Similarly, the Tietze theorem asserts the existence of a factorization of presentation mappings; it is not given as a statement about presentations of the same group. We feel that the style which we have adopted is not only conceptually simpler but also corresponds more exactly to the actual computation of examples in knot theory. The alternative approach, however, is easily described: Consider a presentation

$$(\mathbf{x} : \mathbf{r})_\phi$$

of a group G. The Alexander matrix is defined to be the matrix $\| a_{ij} \|$ of elements

$$a_{ij} = \mathfrak{a}\phi\left(\frac{\partial r_i}{\partial x_j}\right)$$

($\mathfrak{a}$ is, of course, the abelianizer of G), and the elementary ideals are those of this matrix. The invariance theorem then becomes

(4.6) *The elementary ideals are invariants of any finitely presented group* G, i.e., *any two finite presentations of* G *have the same chain of ideals.*

A proof of (4.6) from (4.5) is a simple exercise involving only the most basic properties of homomorphisms and factor groups.

EXERCISES

1. In the group ring of the free group $F(x,y,z)$ calculate the following derivatives:
(a) $\frac{\partial}{\partial x}(xyz^2x^{-1}y^{-1}z^{-2})$, (b) $\frac{\partial}{\partial z}(xyz^2x^{-1}y^{-1}z^{-2})$, (c) $\frac{\partial}{\partial x}[x^m,y]$, (d) $\frac{\partial}{\partial x}[x,y]^n$,
(e) $\frac{\partial}{\partial x}[x^m,y]^n$, (f) $\frac{\partial}{\partial y}[[x,y^m],y^n]$.

2. Prove that if F is a free group with free basis $(x_1, x_2, \cdots)$ and if there exist finitely many elements $a_1, \cdots, a_n$ in JF such that $\sum_{i=1}^{n} a_i(x_i - 1) = 0$, then $a_1 = \cdots = a_n = 0$.

3. If $g_1, g_2, \cdots$ generate G, show directly that for any $\nu \in JG$ there exists a finite set of elements $\nu_1, \cdots, \nu_n$ in JG such that $\nu - \mathfrak{t}\nu = \sum_{i=1}^{n} \nu_i(g_{j_i} - 1)$. (For example, $g_1g_2 - 1 = g_1(g_2 - 1) + (g_1 - 1)$.)

4. One cannot help feeling that our proof of Theorem (2.9) is not quite as good as it might be because it relies essentially on the structure of the free group as it is derived from the semi- group of words. Is it not, after all, possible to give an elegant proof of (2.9) based on the definition of a free group as a group that has a free basis? For example a possible approach would be to prove Exercise 2 above independently of the existence of $\partial/\partial x_j$ and then to combine this result with Exercise 3. Thus from Exercise 3 we get $f(x) - f(1) = \sum_j a_j(x_j - 1)$ for some $a_j \in JF$. If one knew that the elements a_j were uniquely determined by $f(x)$, one could simply define $\frac{\partial f}{\partial x_j}$ to be a_j. Wrestle with this a bit!

5. Using only (2.1), (2.2), and (2.9), prove that if the generator x_j does not occur in the free polynomial $f(x)$, then $\frac{\partial f}{\partial x_j} = 0$. Note that this is a problem about elements of a free group, not about words.

6. Calculate the derivatives of $w = [[a,b],[c,d]]$. What would be the effect on the Alexander matrix of adjoining to a presentation the relation $w = 1$?

7. Denoting by G' the commutator subgroup of a group G and therefore by G'' the commutator subgroup of G', discuss the relationship between the Alexander matrices of G and of G/G''.

8. By constructing the chain of elementary ideals, give another proof of the fact that the free groups of distinct finite ranks m and n are not isomorphic.

9. Suppose θ maps a group G_1 homomorphically onto a group G_2 in such a way that its kernel is contained in the commutator subgroup G_1' of G_1. Prove that the induced homomorphism θ_* of G_1/G_1' onto G_2/G_2' is an isomorphism, and that, for each d, $E_d(G_1)$ is contained in $E_d(G_2)$. ($E_d(G)$ is the dth ideal of the Alexander matrix of any of the presentations of G.)

10. Calculate the chain of elementary ideals for the free abelian group of rank n, and conclude that, for $n > 1$, the free group of rank n is not abelian.

11. Calculate the chain of elementary ideals for:
(a) The fundamental group of a Klein bottle $| a,b : aba^{-1}b = 1 |$;
(b) The group $| a,b : b^2 = 1 |$;
(c) The group $| a,b : b^2 = 1, ab = ba |$;
and use the result to show that these groups are not isomorphic.

12. Calculate the chain of elementary ideals for the metacyclic group $| a,x : a^p = 1, x^{p-1} = 1, xax^{-1} = a^k |$, where p is an odd prime and k is a primitive root modulo p (see the index for a definition of this term). Deduce that this group is not abelian.

13. Calculate the chain of elementary ideals for the fundamental group of

the orientable surface of genus h, $| a_1, b_1, \cdots, a_h, b_h : \prod_{i=1}^{h} [a_i, b_i] = 1 |$, and deduce that if $h \geq 2$ this group is neither free nor abelian.

14. Calculate the chain of elementary ideals for $| x,y : (xy)^n = (yx)^n |$, and deduce that, for $n \geq 2$, this group is neither free nor abelian.

15. Calculate the chain of elementary ideals for the group

$$| x,y,z : [[y^{-1},x],z] = [[z^{-1},y],x] = [[x^{-1},z],y] = 1 |,$$

and deduce that this group is neither free nor abelian.

16. Calculate the chain of elementary ideals for the braid groups[1]

$$\left| \begin{array}{ll} \sigma_1, \cdots, \sigma_n : \sigma_i\sigma_{i+1}\sigma_i = \sigma_{i+1}\sigma_i\sigma_{i+1} & (i = 1, \cdots, n-1), \\ \sigma_i\sigma_j = \sigma_j\sigma_i, & | i - j | \neq 1 \end{array} \right| .$$

17. Prove that if the free group of rank n can be mapped homomorphically onto a group G then $E_n(G) = (1)$.

18. Given any finite set of integral polynomials $f_1(t), \cdots, f_n(t)$ such that $(f_1(1), \cdots, f_n(1)) = 1$, construct a group G such that $G/G' = (t:)$ and $E_1(G) = (f_1(t), \cdots, f_n(t))$.

---

[1] E. Artin, "The Theory of Braids," *American Scientist*, Vol. 38, No. 1 (1950), pp. 112–119; F. Bohnenblust, "The Algebraical Braid Group," *Annals of Mathematics*, Vol. 48 (1947), pp. 127–136.

CHAPTER VIII

The Knot Polynomials

Introduction. The underlying knot-theoretic structure developed in this book is a chain of successively weaker invariants of knot type. The sequence of knot polynomials, to which this chapter is devoted, is the last in the chain

$$\begin{gathered}\text{knot type of } K\\ \downarrow\\ \text{presentation type of } \pi(R^3 - K)\\ \downarrow\\ \text{sequence of elementary ideals}\\ \downarrow\\ \text{sequence of knot polynomials.}\end{gathered}$$

The only complete invariant is the first, i.e., the knot type itself. It is complete for the not very profound reason that two knots are of the same type if and only if they are of the same type. If we stop here, we have a definition but no theory. For all we know, all knots are equivalent. The next step is the major advance. The theorem that knots of the same type possess isomorphic groups reduces the topological problem to a purely algebraic one. The remaining invariants are aimed at the very difficult problem of recognizing when two presentations present nonisomorphic groups. It is important to realize that at each step in the chain information is lost. In fact, for each invariant, we have given at the end of this chapter a pair of knots whose type is distinguished by that invariant but not by the succeeding one. The compensating gain is in the decision problem, i.e., the question of recognizing whether two values of an invariant are the same or different. As we have remarked elsewhere, this problem is unsolvable for group presentations in general. For the knot polynomials, however, it is a triviality.

The knot polynomials can be defined in terms of the elementary ideals (cf. (3.2)). Unlike the ideals, however, which are defined for all finite presentations, the knot polynomials depend for their existence and uniqueness on the special algebraic properties of the abelianized group of a knot group. For this reason, the first section of this chapter is devoted to proving the theorem that the abelianized group of any knot group is infinite cyclic. The second section establishes the necessary algebraic properties of the group ring of an infinite cyclic group. We then define the knot polynomials, check their existence, uniqueness, and invariance, and study some of their properties. The final section contains examples of different knot types distinguished by calculation of their polynomials and ideals. It should be emphasized that we restrict ourselves throughout to knots whose groups we know possess over presentations. By a *knot group*, therefore, we mean now a fundamental group of the complement of a tame knot.

1. The abelianized knot group. Our contention is that the abelianized group of every knot group is infinite cyclic. The proof is based on consideration of the over presentation of a knot group.

Let G be a knot group and $(x_1, \cdots, x_n : r_1, \cdots, r_n)_\phi$ an over presentation of G. A typical example of how a relator r_i is derived by reading around an underpass is shown below in Figure 45. (It obviously doesn't really matter, but

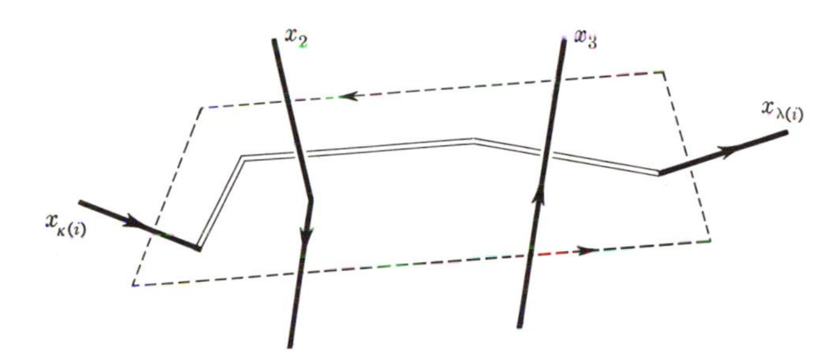

Figure 45. $r_i = x_{\kappa(i)} x_2^{-1} x_3 x_{\lambda(i)}^{-1} x_3^{-1} x_2$

for simplicity we have used the modified over presentation (3.1), Chapter VI, instead of (1.1), Chapter VI.) For any element u in the free group F generated by $x_1, \cdots, x_n$, we define the jth *exponent sum* of u to be the sum of the exponents of x_j at all occurrences of x_j in u. We already have an expression for this quantity; the jth exponent sum of u is the image of $\dfrac{\partial u}{\partial x_j}$ under the trivializer $\mathfrak{t}: JF \to J$. Let $x_{\kappa(i)}$ and $x_{\lambda(i)}$ be the generators corresponding to the two overpasses adjacent to the underpass with respect to which r_i is defined. To be specific, we assume that, with respect to the orientation of the knot, the overpass corresponding to $x_{\kappa(i)}$ precedes the overpass corresponding to $x_{\lambda(i)}$. The example shown in Figure 45 illustrates the fact that the $\kappa(i)$th and $\lambda(i)$th exponent sums of r_i are respectively $+1$ and -1 whereas the exponent sum of r_i with respect to any other generator is 0. Hence, if θ is any homomorphism of G into an abelian group, we have

$$1 = \theta\phi r_i = (\theta\phi x_{\kappa(i)})(\theta\phi x_{\lambda(i)})^{-1}.$$

Since every knot projection is connected, we conclude that $\theta\phi x_i = \theta\phi x_j$ for every pair of generators x_i and x_j. Thus any element in the image group θG is a power of the single element $t = \theta\phi x_j$, $j = 1, \cdots, n$. We have proved

(1.1) *Every abelian homomorphic image of a knot group is cyclic. Furthermore, the generators of any over presentation are all mapped onto a single generator.*

In particular, the abelianized group of any knot group is cyclic. It remains to prove that it cannot be finite. To prove this, consider again the over presentation of G, and denote by (t) an infinite cyclic group generated by t. Since F is a free group, the assignment $\zeta x_j = t$, $j = 1, \cdots, n$ can be extended to a homomorphism of F onto (t). It is easy to show that there exists a homomorphism θ of G onto (t) such that the diagram

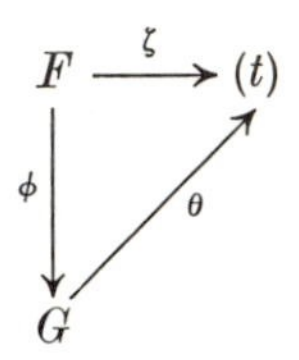

is consistent. For, clearly,

$$\zeta r_i = t^{s_i}, \qquad i = 1, \cdots, n,$$

where s_i is the sum over j of the exponent sums of r_i with respect to x_j. We have

$$s_i = \sum_{j=1}^{n} \mathfrak{t} \frac{\partial r_i}{\partial x_j} = \sum_{j=1}^{n} (\delta_{j,\kappa(i)} - \delta_{j,\lambda(i)}) = 0.$$

Thus, $\zeta r_i = 1$, $i = 1, \cdots, n$, and the consequence of $r_i, \cdots, \hat{r}_k, \cdots, r_n$, which is the kernel of ϕ, is contained in the kernel of ζ. It follows that θ is well-defined by

$$\theta\phi u = \zeta u, \qquad u \in F.$$

Since ζ is onto, so is θ. Consider next the abelianizer $\mathfrak{a}\colon G \to G/[G,G]$. We recall the important fact that any homomorphism of a group into an abelian group can be factored through the commutator quotient group (cf. (4.4), Chapter IV). As a result, there exists a homomorphism θ' such that the diagram

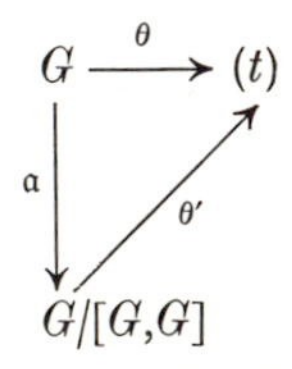

is consistent. Since θ is onto, so is θ'. A function whose image is infinite cannot have a finite domain; so we conclude that $G/[G,G]$ is infinite. Combining this result with (1.1), we obtain

(1.2) *The abelianized group of any knot group is infinite cyclic.*

Another way of arriving at (1.2) is by reducing a presentation by Tietze operations. If one begins, for example, with a Wirtinger presentation of G, then $G/[G,G]$ has the presentation (cf. (4.6), Chapter IV)

$$(x_1, \cdots, x_n \ : \ x_{i+1}^{-1} x_{k(i)}^{\epsilon} x_i x_{k(i)}^{-\epsilon}, [x_i, x_j], i, j = 1, \cdots, n),$$

which one can then reduce to

$$(x_1, \cdots, x_n \ : \ x_{i+1} x_i^{-1}, \quad i = 1, \cdots, n) \simeq (t \ :).$$

Notice that this approach is based on the theorem (cf. (4.5), Chapter IV) that the consequence of the set of commutators $[g_i, g_j]$, where $g_1, g_2, \cdots$ generate a group, is the whole commutator subgroup.

2. The group ring of an infinite cyclic group. A knowledge of some of the basic algebraic properties of the group ring of an infinite cyclic group H is necessary for a proper understanding of the knot polynomials. In this section, therefore, we shall review some of the elementary concepts of divisibility in rings and integral domains in general and see how they apply to the group ring JH.

Let R be an arbitrary ring having a multiplicative identity 1. An element u of R is called a *unit* if it has a left and a right inverse, i.e., if there exist $v, w \in R$ such that $uv = wu = 1$. The associative law implies that

$$w = w(uv) = (wu)v = v.$$

Hence, an equivalent definition is that a unit is an element having an inverse, which, by the same reasoning, must be unique. Since the product of any two units is again a unit, it is easy to see that *the set of units of R is a multiplicative group*. For example, the only units of the ring of integers are $+1$ and -1. In a group ring, all group elements and their negatives are obviously units. They are the so-called *trivial units* of the group ring. The possibility of the existence of nontrivial units will be considered briefly somewhat later in this section.

For any elements a and b of a commutative ring R, we say that *a divides b*, written $a \mid b$, if there exists $c \in R$ such that $b = ac$. Elements a and b are *associates* if $a \mid b$ and $b \mid a$. This relation is an equivalence relation provided R contains an identity 1. The only associate of 0 is 0 itself. A commutative ring is called an *integral domain* if it contains at least two elements and has the property that if $a \neq 0$ and $b \neq 0$, then $ab \neq 0$.

(2.1) *Two elements in an integral domain with identity* 1 *are associates if and only if one is a unit multiple of the other.*

Proof. If a and b are associates, there exist elements c and d such that $a = bc$ and $b = ad$. Consequently, $a = adc$ and

$$a(1 - dc) = 0.$$

Since the ring is an integral domain, either a or $(1 - dc)$ is 0. If $a = 0$ it must also be true that $b = 0$, and we are done. If $1 - dc = 0$, then d and c are units, and the desired conclusion again follows. Conversely, suppose $a = ub$ for some unit u. Then, $b = u^{-1}a$ and a and b are associates, and we are finished.

A commutative ring R will be said to be *associate to a subring* Q of R if there exists a mapping $\rho\colon R \to Q$ such that, for any $a, b \in R$, the elements a and ρa are associates, and $\rho(ab) = (\rho a)(\rho b)$. It is immediate that

(2.2) *If Q is an integral domain, then so is R.*

Proof. Consider nonzero elements a and b in R. Any associate of a nonzero element must also be nonzero; hence ρa and ρb are nonzero. Since Q is an integral domain, $\rho(ab) = \rho a \rho b \neq 0$, and therefore $ab \neq 0$.

An element d of a commutative ring R is called a *greatest common divisor*, abbreviated g.c.d., of a finite set of elements $a_1, \cdots, a_n \in R$ if $d \mid a_i$, $i = 1, \cdots, n$, and, for any $e \in R$, if $e \mid a_i$, $i = 1, \cdots, n$, then $e \mid d$. Obviously, any two g.c.d.'s of the same finite set are associates. There is no reason for supposing that every finite set of elements in an arbitrary commutative ring has a g.c.d. A counter-example to this conjecture is provided by the ring of all complex numbers $m + n\sqrt{-3}$ where m and n are integers. Since it is a nontrivial subring of the field of complex numbers, this ring is automatically an integral domain. The only units are 1 and -1. It is not difficult to show that any common divisor of 4 and $2(1 + \sqrt{-3})$ is one of the numbers 1, 2, $1 + \sqrt{-3}$, $1 - \sqrt{-3}$, and their negatives. Since no one of these numbers is divisible by all of the others, it follows that 4 and $2(1 + \sqrt{-3})$ have no greatest common divisor. A ring will be called a *g.c.d. domain* if it is an integral domain and every finite set of elements has a g.c.d.

(2.3) *If a commutative ring R is associate to a subring Q which is a g.c.d. domain, then R is also a g.c.d. domain.*

Proof. It follows from (2.2) that R is an integral domain. The product-preserving mapping of R into Q is denoted, as before, by $\rho\colon R \to Q$. Consider any finite set of elements $a_1, \cdots, a_n$ in R, and let d be a g.c.d. in Q of $\rho a_1, \cdots, \rho a_n$. We contend that d is a g.c.d. in R of $a_1, \cdots, a_n$. First of all, since $d \mid \rho a_i$ and $\rho a_i \mid a_i$, we have $d \mid a_i$, $i = 1, \cdots, n$. Next suppose $e \mid a_i$, $i = 1, \cdots, n$. Then, $a_i = b_i e$ and $\rho a_i = \rho b_i \rho e$; so ρe divides ρa_i in Q. Since d is a g.c.d., ρe divides d in Q and therefore also in R. Thus, $e \mid \rho e$ and $\rho e \mid d$, and so $e \mid d$. This completes the proof.

In an integral domain with identity, any two g.c.d.'s of the same finite set are unit multiples of each other. As a result, in such a ring it is customary to speak of any greatest common divisor d of a finite set $a_1, \cdots, a_n$ as *the greatest common divisor* and to write $d = \text{g.c.d.}(a_1, \cdots, a_n)$.

The concepts of unique factorization and of a prime are also relevant to the present discussion. An element p in an integral domain with identity is a *prime* if p is not a unit and if $p = ab$ implies that either a or b is a unit. A *unique factorization domain* is an integral domain with identity in which every element which is neither zero nor a unit has an essentially unique factorization into primes. To say that factorization into primes is essentially unique means that, for any primes p_i, q_j, $i = 1, \cdots, m$, $j = 1, \cdots, n$, if $p_1 \cdots p_m = q_1 \cdots q_n$, then $m = n$ and, for a suitable ordering, p_i and q_i are associates, $i = 1, \cdots, n$. The statement that the ring of integers is a unique factorization domain is just the famous Fundamental Theorem of Arithmetic.

(2.4) *Every unique factorization domain is a g.c.d. domain.*

Proof. If a is any nonzero element of a unique factorization domain R, the primes which are associates may be combined and a factorization

$$a = up_1^{n_1} \cdots p_m^{n_m}, \qquad m \geq 0,$$

obtained in which no two of $p_1, \cdots, p_m$ are associates, the n_j are positive integers, and u is a unit. Any divisor of a has a factorization $u'p_1^{n_1'} \cdots p_m^{n_m'}$ where u' is a unit and the n_j' are integers such that $0 \leq n_j' \leq n_j$. Similarly, if $a_1, \cdots, a_n$ are nonzero elements of R, there exist primes $p_1, \cdots, p_m$, $m \geq 0$, no two of which are associates, such that

$$a_i = u_i p_1^{n_{i1}} \cdots p_m^{n_{im}}, \qquad i = 1, \cdots, n,$$

where the n_{ij} are integers ≥ 0, and the u_i are units. The element

$$d = p_1^{n_1} \cdots p_m^{n_m}, \qquad n_j = \min_i (n_{ij}),$$

is obviously a greatest common divisor of $a_1, \cdots, a_n$. An integral domain in which every finite set of nonzero elements has a g.c.d. is a g.c.d. domain; so the proof is complete.

We assume that the reader has some familiarity with the definition and elementary facts about the ring $R[t]$ of polynomials in one variable t with coefficients in an integral domain R. For example,[1] it is easy to show that *if R is an integral domain, then $R[t]$ is an integral domain whose only units are the units of R.* A deeper result, the key point in the proof of which is due to Gauss, is the following theorem.[2]

(2.5) *If R is a unique factorization domain, then so is $R[t]$.*

---

[1] See N. Jacobson, *Lectures in Abstract Algebra*, Vol. 1 (D. van Nostrand Company, Inc.; Princeton, N.J., 1951), Chap. 3, Sects. 4, 5, 6.

[2] See N. Jacobson, *Lectures in Abstract Algebra*, Vol. 1 (D. van Nostrand Company, Inc.; Princeton, N.J., 1951), Chap. 4, Sect. 6.

We return now to the group ring of the infinite cyclic group H. For a choice of a generator t of H, an arbitrary element a of JH has a unique representation

$$a = \sum_{-\infty}^{\infty} a_n t^n,$$

where all but a finite number of integers a_n are equal to zero. It follows that the polynomial ring $J[t]$ is a subring of JH. For every nonzero element a in JH, we define $\mu(a)$ to be the smallest integer n such that $a_n \neq 0$. For example,

$$\mu(t^3 + 2t - 7t^{-5}) = -5, \qquad \mu(1) = 0.$$

If $a = 0$, we set $\mu(a) = \infty$ with the usual convention that $\infty + \infty = \infty$ and $\infty + n = n + \infty = \infty$. Then,

(2.6) $\mu(ab) = \mu(a) + \mu(b)$, *for* $a, b \in JH$.

Proof. If either $a = 0$ or $b = 0$, our convention gives the result; so we assume that both are nonzero. Let $c = ab$. Then, if

$$a = \sum_{-\infty}^{\infty} a_n t^n, \quad b = \sum_{-\infty}^{\infty} b_n t^n, \quad \text{and} \quad c = \sum_{-\infty}^{\infty} c_n t^n,$$

we have

$$c_n = \sum_{i=-\infty}^{\infty} a_i b_{n-i} = \sum_{i=\mu(a)}^{\infty} a_i b_{n-i}.$$

If $n < \mu(a) + \mu(b)$ and $i \geq \mu(a)$, then $n - i \leq n - \mu(a) < \mu(b)$ and so $a_i b_{n-i} = 0$. Hence,

$$c_n = 0 \quad \text{for } n < \mu(a) + \mu(b).$$

If $n = \mu(a) + \mu(b)$ and $i > \mu(a)$, then $n - i < n - \mu(a) = \mu(b)$ and so $a_i b_{n-i} = 0$. Thus,

$$c_{\mu(a)+\mu(b)} = a_{\mu(a)} b_{\mu(b)} \neq 0$$

because the ring of integers is an integral domain. This completes the proof.

With the convention that $t^{-\infty} = 0$, it is apparent that, for any $a \in JH$, the element $at^{-\mu(a)}$ is a polynomial. Hence, the function given by $\rho a = at^{-\mu(a)}$ defines a mapping of JH into the subring $J[t]$ of polynomials. The fact that any power of t is a unit of JH implies that, for any $a \in JH$, a and ρa are associates. Since $t^{-\mu(a)}t^{-\mu(b)} = t^{-(\mu(a)+\mu(b))}$ for any $a, b \in JH$, it is a corollary of (2.6) that ρ is product-preserving. We conclude that JH is associate to the subring $J[t]$ of polynomials. As a consequence of the Fundamental Theorem of Arithmetic and Lemmas (2.5) and (2.4), it follows that $J[t]$ is a g.c.d. domain. Hence, by (2.3), we obtain our main theorem:

(2.7) *The group ring of an infinite cyclic group is a g.c.d. domain.*

It is worth noting that JH is an integral domain as a trivial consequence of (2.6). An example of a group ring which is not an integral domain is the group ring of a cyclic group of order 2, in which

$$(t+1)(t-1) = t^2 - 1 = 0$$

is a valid equation. Another important result is

(2.8) *The group ring of an infinite cyclic group has only trivial units, i.e., the powers of a generator t and their negatives.*

Proof. Let a be a unit of JH and b its inverse. Then, $ab = 1$ and

$$(\rho a)(\rho b) = \rho 1 = 1.$$

The only units of the polynomial ring $J[t]$ are 1 and -1. Hence,

$$\rho a = at^{-\mu(a)} = \pm 1,$$
$$a = \pm t^{\mu(a)},$$

and the proof is complete.

For an example of nontrivial units in a group ring, consider the group ring of the cyclic group of order 5 generated by t. Then,

$$\begin{aligned}(1 - t^2 + t^4)(1 - t + t^2)t^2 &= (1 - t + t^3 - t^5 + t^6)t^2 \\ &= (1 - t + t^3 - 1 + t)t^2 \\ &= t^5 = 1.\end{aligned}$$

Our conclusions about the group ring of an infinite cyclic group are actually valid for the group ring of any finitely generated free abelian group. The proof is the same. Let K be a free abelian (multiplicative) group of rank m generated by $t_1, \cdots, t_m$. Then, an arbitrary element a of JK has a unique representation

$$a = \sum_{n_1, \cdots, n_m = -\infty}^{\infty} a_{n_1, \cdots, n_m} t_1^{n_1} \cdots t_m^{n_m},$$

where all but a finite number of the integers $a_{n_1, \cdots, n_m}$ are zero. For $a \neq 0$, we define $\mu_i(a)$ to be the smallest power of t_i which occurs in a. The function ρ given by

$$\rho a = \begin{cases} at_1^{-\mu_1(a)} \cdots t_m^{-\mu_m(a)}, & a \neq 0, \\ 0, & a = 0, \end{cases}$$

defines a product-preserving mapping of JK into the polynomial ring in m variables $J[t_1, \cdots, t_m]$. Clearly, every element is associate to its image under ρ, and so JK is associate to $J[t_1, \cdots, t_m]$. Starting with the ring of integers, we obtain after m applications of Theorem (2.5) the fact that $J[t_1, \cdots, t_m]$ is a unique factorization domain, and a similar argument shows that the only units are 1 and -1. Lemmas (2.4) and (2.3) together with the obvious analogue of (2.8) then complete the proof of

(2.9) *The group ring of a free abelian group of rank $m \geq 0$ is a g.c.d. domain whose only units are group elements and their negatives.*

The notion of a greatest common divisor can be elegantly described in terms of ideals. We recall that, in a commutative ring R with identity 1, the *ideal E generated by a subset S* is the set of all finite sums

$$\sum_i a_i b_i, \qquad a_i \in S, b_i \in R.$$

An equivalent characterization of E is as the smallest ideal containing S. By this is meant the intersection of all ideals of R which contain S. Since any ideal which contains S must contain the ideal generated by S, the equivalence of the two characterizations is obvious. An ideal is called a *principal ideal* if it is generated by a single element. It is a simple matter to check that *any two associates generate the same ideal in R, and, conversely, any two generators of a principal ideal of R are associates.* A notion we shall find useful is that of the smallest principal ideal containing a given finite set of elements. The only trouble is that, for commutative rings in general, there isn't any such thing. The smallest principal ideal containing $a_1, \cdots, a_n$ means the intersection of all principal ideals of the given ring which contain $a_1, \cdots, a_n$. This intersection is an ideal all right; but it doesn't have to be a principal ideal.

(2.10) *If R is a commutative ring with identity and $a_1, \cdots, a_n \in R$, then d is a g.c.d. of $a_1, \cdots, a_n$ if and only if the intersection of all principal ideals of R which contain $a_1, \cdots, a_n$ is itself a principal ideal generated by d.*

Proof. Suppose first that the smallest principal ideal D containing $a_1, \cdots, a_n$ does exist and that d is a generator. Then we certainly have $d \mid a_i$, $i = 1, \cdots, n$. Consider next an arbitrary $e \in R$ such that $e \mid a_i$, $i = 1, \cdots, n$. The principal ideal E generated by e contains $a_1, \cdots, a_n$ and, therefore, also contains D. Thus $d \in E$ and so $e \mid d$. We conclude that d is a g.c.d. of $a_1, \cdots, a_n$. Conversely, suppose d is given as a g.c.d. of $a_1, \cdots, a_n$ and generates the principal ideal D. Since $d \mid a_i$, $i = 1, \cdots, n$, it follows that D contains $a_1, \cdots, a_n$. Consider next any principal ideal E containing $a_1, \cdots, a_n$. If e is a generator of E, we have $e \mid a_i, i = 1, \cdots, n$. Consequently, $e \mid d$ and this implies that E contains D. Any member of a collection of sets which is a subset of every set in the collection must itself equal the intersection of the collection. It follows that D is the smallest principal ideal containing $a_1, \cdots, a_n$, and the proof of (2.10) is complete. As a corollary, we have

(2.11) *In a g.c.d. domain with identity, the g.c.d. of any finite set of elements is the generator of the smallest principal ideal that contains them.*

The generator of a principal ideal in such a ring is, of course, determined only to within unit multiples, and the same goes for the g.c.d. Thus this last result is true insofar as it makes sense; strictly speaking, it is the equivalence classes which are equal.

3. The knot polynomials. The group ring of an infinite cyclic group becomes, upon selection of a generator t of the group, what may be called a ring of L-*polynomials* in t. The letter "L" is suggested by the Laurent power series with negative exponents which arises in the theory of complex variables. More generally, a ring of L-polynomials in n variables $t_1, \cdots, t_n$ is the group ring of a free abelian group of rank n generated by $t_1, \cdots, t_n$. Notice that, for one variable, the notion of L-polynomial coincides with that of free polynomial which was introduced in the last chapter. This is simply because an infinite cyclic group happens to be both free and free abelian. It should be emphasized that the ring of L-polynomials is not determined by the group ring alone. An element of the group ring of an infinite cyclic group generally has two representations as an L-polynomial, e.g.,

$$3t^2 - 5t + t^{-3} \quad \text{and} \quad t^3 - 5t^{-1} + 3t^{-2},$$

depending on which of the two generators is set equal to t. Nevertheless, we commonly refer to an element of the group ring of a free abelian group as an L-polynomial. In fact, the following definition of the knot polynomials is an example of this practice.

For any integer $k \geq 0$, the kth *knot polynomial* Δ_k of a finite presentation $(\mathbf{x} : \mathbf{r}) = (x_1, \cdots, x_n : r_1, \cdots, r_m)$ of a knot group is the g.c.d. of the determinants of all $(n - k) \times (n - k)$ submatrices of the Alexander matrix of $(\mathbf{x} : \mathbf{r})$ where it is understood that

$$\Delta_k = 0 \quad \text{if} \quad n - k > m,$$
$$\Delta_k = 1 \quad \text{if} \quad n - k \leq 0.$$

The group $|\mathbf{x} : \mathbf{r}|$ is canonically isomorphic to the knot group it presents; hence, by (1.2), the abelianized group is certainly infinite cyclic. It follows from (2.7) and (2.8) that the group ring of the abelianized group of $|\mathbf{x} : \mathbf{r}|$ is a g.c.d. domain with only trivial units. We conclude that

(3.1) *The knot polynomials exist and are unique to within* $\pm t^n$, *where* n *is any integer and* t *is a generator of the infinite cyclic abelianized group of the presentation* $(\mathbf{x} : \mathbf{r})$ *of the knot group.*

The smallest principal ideal containing a given finite set of elements is the smallest principal ideal containing the ideal generated by this finite set of elements. Hence, as a consequence of (2.11) and the definitions of the polynomial Δ_k and the elementary ideal E_k, we obtain the following characterization of the knot polynomials.

(3.2) *Each knot polynomial* Δ_k *is the generator of the smallest principal ideal containing the elementary ideal* E_k.

A very important practical corollary of (3.2) and our result that equivalent matrices have the same elementary ideals is the fact that the knot polynomials of a presentation, like the elementary ideals, can be calculated from any matrix equivalent to the Alexander matrix. One naturally uses the simplest matrix one can find.

(3.3) $\Delta_{k+1} \mid \Delta_k$.

Proof. We shall use a corollary of (3.2) and the fact that the elementary ideals form an ascending chain (cf. (4.1), Chapter VII). Let (Δ_k) and (Δ_{k+1}) denote the principal ideals generated by Δ_k and Δ_{k+1}, respectively. We have

$$(\Delta_{k+1}) \supset E_{k+1} \supset E_k.$$

Since (Δ_k) is the smallest principal ideal containing E_k,

$$(\Delta_{k+1}) \supset (\Delta_k).$$

Thus, $\Delta_k = a\Delta_{k+1}$, or $\Delta_{k+1} \mid \Delta_k$.

The next theorem is the analogue for knot groups of the invariance theorem for the elementary ideals, (4.5) of Chapter VII. Its essential content is that *the knot polynomials are invariants of knot type.*

(3.4) Invariance of the knot polynomials. *If* $(\mathbf{x} : \mathbf{r})$ *and* $(\mathbf{y} : \mathbf{s})$ *are finite presentations of knot groups and*

$$f\colon (\mathbf{x} : \mathbf{r}) \to (\mathbf{y} : \mathbf{s})$$

is a presentation equivalence, then, to within units, the kth knot polynomial Δ_k *of* $(\mathbf{x} : \mathbf{r})$ *is mapped by* f_{**} *onto the kth knot polynomial* Δ_k' *of* $(\mathbf{y} : \mathbf{s})$.

Proof. We shall use a corollary of (3.2) and the Invariance Theorem (4.5) of Chapter VII. We recall that f_{**} is the linear extension to the group rings of an induced isomorphism of the abelianized group of $| \mathbf{x} : \mathbf{r} |$ onto that of $| \mathbf{y} : \mathbf{s} |$ (cf. (4.4) and preceding paragraph in Chapter VII). Denote by (Δ_k) and (Δ_k') the principal ideals generated by Δ_k and Δ_k', respectively, and by E_k and E_k' the elementary ideals of $(\mathbf{x} : \mathbf{r})$ and $(\mathbf{y} : \mathbf{s})$, respectively. Then,

$$E_k \subset (\Delta_k) \quad \text{and} \quad E_k' \subset (\Delta_k') \quad \text{and} \quad f_{**}E_k = E_k'.$$

Now, an isomorphic image of a principal ideal is principal, and $f_{**}(\Delta_k) \supset f_{**}E_k = E_k'$. Since (Δ_k') is minimal,

$$f_{**}(\Delta_k) \supset (\Delta_k').$$

By the same argument,

$$f_{**}^{-1}(\Delta_k') \supset (\Delta_k),$$

and so

$$f_{**}(\Delta_k) = (\Delta_k').$$

Since $f_{**}(\Delta_k)$ is generated by $f_{**}\Delta_k$, the elements Δ_k' and $f_{**}\Delta_k$ are associates and therefore unit multiples of each other. This completes the proof.

The preceding theorem is of fundamental importance. It implies directly that the knot polynomials are invariants of knot type. We recall the basic principles: (1) If two knots represent the same knot type, their groups are isomorphic (cf. (4.7) and the subsequent discussion in Chapter II). (2) Two presentations of isomorphic groups are of the same presentation type, and, hence, there exists a presentation equivalence between them (cf. (2.4) and accompanying discussion in Chapter IV). Let us see how this invariance theorem applies to a fictitious example. Suppose we are given two knots K and K', known to be of the same type, and are asked to obtain, for some integer k, their respective polynomials Δ_k and Δ_k'. We determine an over presentation of K, possibly simplify it, calculate the matrix of derivatives, select a generator t of the abelianized group, and obtain the Alexander matrix. We then manage to find a g.c.d. of all determinants of order $n - k$ and, finally, end up with an L-polynomial, say $\Delta_k = 3t^3 - 5t^2 + t$. Since the knot polynomials are unique only up to units, it is natural to *normalize* Δ_k to the polynomial

$$\Delta_k = 3t^2 - 5t + 1,$$

i.e., no negative powers of t and a positive constant term. Notice, however, that the form of the normalized polynomial Δ_k depends on the choice of generator of the abelianized group of the presentation. For if we select the other generator $s = t^{-1}$, the final normalized polynomial is

$$\Delta_k = s^2 - 5s + 3,$$

which is not of the same form as $3t^2 - 5t + 1$ but is, in the ring JH, an associate of it. The next question is: If we go through a similar calculation for K' and obtain Δ_k', what must it look like? Let us assume that we have picked a generator x so that Δ_k' is an L-polynomial in x. Since K and K' are of the same type, their knot groups are isomorphic and there exists an isomorphism f_{**} between the abelianized groups of the presentations of the knot groups. Notice that regardless of how difficult it may be to describe the isomorphism between the knot groups, the mapping f_{**} is very simple. There are only two ways to map one infinite cyclic group isomorphically onto another: either $f_{**}(t) = x$ or $f_{**}(t) = x^{-1}$. In our example, therefore, either

$$f_{**}\Delta_k = 3x^2 - 5x + 1 \quad \text{or} \quad 3x^{-2} - 5x^{-1} + 1.$$

By the Invariance Theorem (3.4), we have $f_{**}\Delta_k = \pm x^n \Delta_k'$. Hence, if Δ_k' is normalized, there are just the two possibilities

$$\Delta_k' = 3x^2 - 5x + 1 \quad \text{or} \quad x^2 - 5x + 3.$$

As we have indicated above, these two reciprocal forms of the polynomial are equally good.

The above fictitious example is very fictitious indeed. We shall prove in the next chapter that if $\Delta_k(t) = c_n t^n + c_{n-1}t^{n-1} + \cdots + c_0$ is an arbitrary normalized knot polynomial, then $c_i = c_{n-i}$, $i = 0, \cdots, n$. As a result of this symmetry, it follows that the normalized form of a knot polynomial is an invariant of knot type. Specifically, if the normalized knot polynomials Δ_k and Δ_k' of respective knots K and K' are not identical for every $k \geq 0$, then K and K' represent distinct knot types.

In any over presentation of a knot group any relator is a consequence of the others and therefore can be dropped (cf. (1.3), Chapter VI). Thus, every knot group has a presentation whose Alexander matrix is $(n - 1) \times n$. Consequently,

(3.5) *The 0th elementary ideal and polynomial of a knot group are trivial*: $E_0 = \Delta_0 = 0$.

This is actually a special case of the more general fact that the 0th elementary ideal of a finite presentation is trivial if and only if the abelianized group of the presentation is infinite.

Let $(\mathbf{x} : \mathbf{r}) = (x_1, \cdots, x_n : r_1, \cdots, r_m)$ be a presentation of a given knot group. We denote by γ the homomorphism of the free group generated by $x_1, \cdots, x_n$ onto the factor group $| \mathbf{x} : \mathbf{r} |$, and by $\mathfrak{a}$ the abelianizer of $| \mathbf{x} : \mathbf{r} |$. Suppose that all the generators of the presentation are mapped by $\mathfrak{a}\gamma$ onto a single element, i.e.,

(3.6) $\mathfrak{a}\gamma x_i = \mathfrak{a}\gamma x_j, \qquad i, j = 1, \cdots, n.$

This condition is in fact satisfied by any over presentation (cf. (1.1)). The entries of the Alexander matrix $A = \| a_{ij} \|$ of $(\mathbf{x} : \mathbf{r})$ are defined by

$$a_{ij} = \mathfrak{a}\gamma\left(\frac{\partial r_i}{\partial x_j}\right), \quad i = 1, \cdots, m \text{ and } j = 1, \cdots, n.$$

By the fundamental formula (cf. (2.11), Chapter VII),

$$r_i - 1 = \sum_{j=1}^{n} \left(\frac{\partial r_i}{\partial x_j}\right) (x_j - 1).$$

Since $\gamma r_i = 1$,

$$0 = \sum_{j=1}^{n} \mathfrak{a}\gamma\left(\frac{\partial r_i}{\partial x_j}\right) (\mathfrak{a}\gamma x_j - 1).$$

Since $\mathfrak{a}\gamma x_1 = \mathfrak{a}\gamma x_j, j = 1, \cdots, n$, we can write

$$0 = (\sum_{j=1}^{n} a_{ij})(\mathfrak{a}\gamma x_1 - 1).$$

The element $\mathfrak{a}\gamma x_1$ is a generator of the infinite cyclic abelianized group of $| \mathbf{x} : \mathbf{r} |$; so $(\mathfrak{a}\gamma x_1 - 1) \neq 0$. Since the group ring of an infinite cyclic group is an integral domain,

$$0 = \sum_{j=1}^{n} a_{ij},$$

i.e., the sum of the column vectors of the Alexander matrix is the zero vector.

Hence,

(3.7) *The Alexander matrix A of any finite presentation of a knot group which satisfies* (3.6) *is equivalent to the matrix obtained by replacing any column of A with a column of zeros.*

Suppose $(\mathbf{x} : \mathbf{r})$ is an over presentation with n generators and $n - 1$ relators. As remarked above, (3.6) holds, and so (3.7) is applicable. In this case, however, A is an $(n - 1) \times n$ matrix. If one column is replaced by zeros, there remains at most one $(n - 1) \times (n - 1)$ submatrix with non-zero determinant. Since equivalent matrices define the same elementary ideals, it follows that the 1st elementary ideal of $(\mathbf{x} : \mathbf{r})$ is principal. Hence, by (3.2),

(3.8) *The* 1*st elementary ideal of a knot group is a principal ideal generated by the* 1*st knot polynomial* Δ_1.

The 1st knot polynomial Δ_1 is the most important member of the sequence of knot polynomials. It is called the *Alexander polynomial* of the knot group and is commonly written without the subscript. Thus,

$$\Delta(t) = \Delta_1(t).$$

It also follows, of course, that the determinant of any one of the $(n - 1) \times (n - 1)$ submatrices of A may be taken to be the polynomial Δ_1.

4. Knot types and knot polynomials. The following examples illustrate the power of the knot polynomials. It will become apparent that these invariants provide a systematic tool for distinguishing knot types on quite a respectable scale. The computational procedure is based on the results of the preceding sections. For example, it is a consequence of (1.1) that the Alexander matrix of an over presentation can be obtained from the matrix of derivatives simply by setting all generators x_i equal to t. On the other hand, it is usually to one's advantage to simplify an over presentation before starting to compute derivatives. However, if all the generators of one presentation of a knot group are mapped by $\mathfrak{a}\gamma$ onto a single generator of the abelianized group, the same will be true for any other presentation obtained from the first by means of Tietze operations **I**, **I′**, and **II′**. So one may still set all generators x_i equal to t after simplifying an over presentation provided no new generators have been introduced in the process. Notice, moreover, that (3.7) is valid for such presentations. This fact obviously offers a substantial computational shortcut.

More often than not, a group presentation is written with relations rather than the more formal relators:

$$G = | x_1, \cdots, x_n : r_1 = s_1, \cdots, r_m = s_m |.$$

The relation $r_i = s_i$ corresponds to the relator $r_i s_i^{-1}$. We have

$$\frac{\partial r_i s_i^{-1}}{\partial x_j} = \frac{\partial r_i}{\partial x_j} - r_i s_i^{-1} \frac{\partial s_i}{\partial x_j}.$$

Since the canonical homomorphism γ maps every relator onto 1, computation of the entries a_{ij} of the Alexander matrix is simplified by the observation that

$$a_{ij} = \mathfrak{a}\gamma\left(\frac{\partial r_i s_i^{-1}}{\partial x_j}\right) = \mathfrak{a}\gamma \frac{\partial}{\partial x_j}(r_i - s_i).$$

In the examples which follow we consider first those knots for which we have already computed group presentations in Chapter VI.

(4.1) *Trivial knot* (Figure 46).

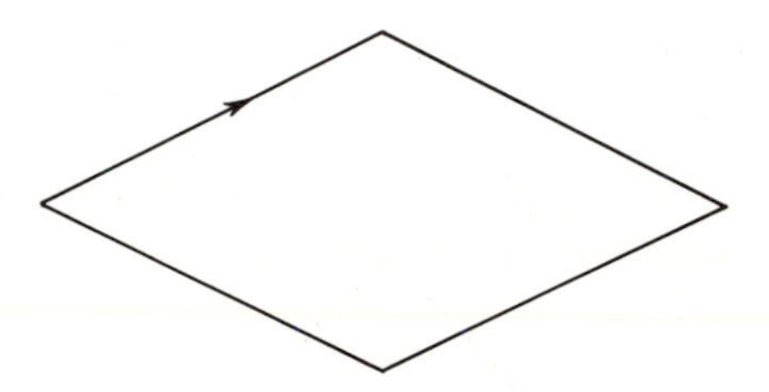

Figure 46. $\pi(R^3 - K) = |\mathbf{x} :|$

Rather than talk about a matrix with one column and no rows, we observe that the presentation $(x\ :)$ is of the same type as $(x\ :\ 1)$. Hence, the Alexander matrix is simply $A = \| 0 \|$ and

$$\Delta_k = 1 \quad \text{for} \quad k \geq 1.$$

(4.2) *Clover-leaf knot* (Figure 47).

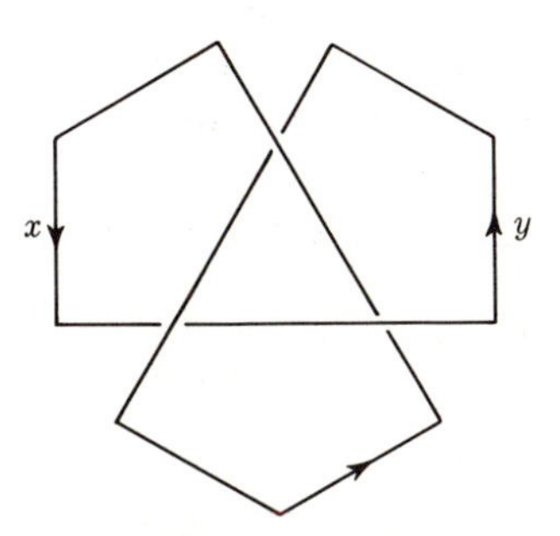

Figure 47. $\pi(R^3 - K) = |\ x,y : xyx = yxy\ |$

The two entries of the Alexander matrix $A = \| a_{11} \ a_{12} \|$ are

$$a_{11} = \mathfrak{a}\gamma \frac{\partial}{\partial x}(xyx - yxy) = \mathfrak{a}\gamma(1 + xy - y),$$

$$a_{12} = \mathfrak{a}\gamma \frac{\partial}{\partial y}(xyx - yxy) = \mathfrak{a}\gamma(x - 1 - yx).$$

Setting $\mathfrak{a}\gamma x = \mathfrak{a}\gamma y = t$, we obtain

$$A = \| 1 - t + t^2 \qquad -1 + t - t^2 \|.$$

Hence,

$$\Delta_1 = 1 - t + t^2 \quad \text{and} \quad \Delta_k = 1 \quad \text{for} \quad k \geq 2.$$

So the clover-leaf knot cannot be untied. We have, however, already proved this fact in Chapter VI.

(4.3) *Figure-eight knot.* A Wirtinger presentation in which x and y correspond to the overpasses shown in Figure 48 can be simplified to give

$$\pi(R^3 - K) = | \, x,y \ : \ yx^{-1}yxy^{-1} = x^{-1}yxy^{-1}x \, |.$$

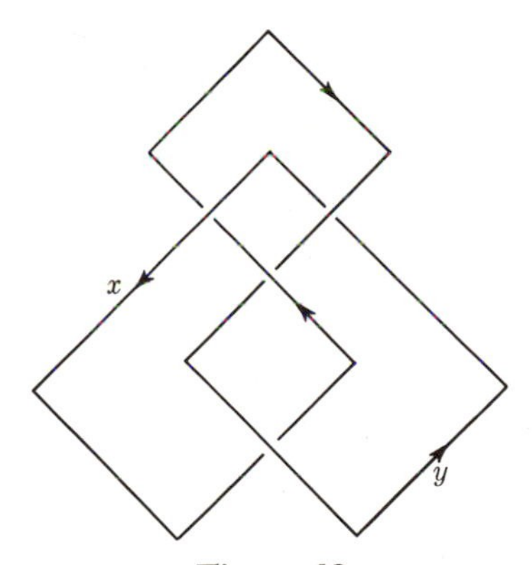

Figure 48

Again, we set $\mathfrak{a}\gamma x = \mathfrak{a}\gamma y = t$. The computation is halved by the observation that since (3.7) holds for the above presentation, either one of the two entries of the Alexander matrix may be set equal to the polynomial Δ_1. Thus

$$\Delta_1 = \mathfrak{a}\gamma \frac{\partial}{\partial x}(yx^{-1}yxy^{-1} - x^{-1}yxy^{-1}x)$$

$$= -1 + t + t^{-1} - 1 - 1 = t - 3 + t^{-1}.$$

Normalizing, we obtain

$$\Delta_1 = t^2 - 3t + 1.$$

Obviously, $\Delta_k = 1$ for $k \geq 2$. We conclude that the figure-eight knot is not trivial and is of a different type from the clover-leaf.

(4.4) *Three-lead four bight Turk's head knot* (Figure 49).

$$\pi(R^3 - K) = | x_1, x_2, x_3, x_4 \ : \ x_i = [x_{i+3}, x_{i+2}^{-1}]x_{i+1}[x_{i+2}^{-1}, x_{i+3}],$$
$$i = 1, \cdots, 4 \text{ integers mod } 4 |.$$

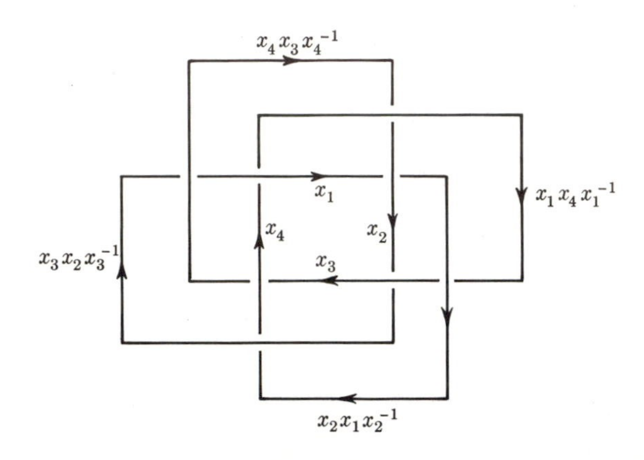

Figure 49

As can be read from Figure 49, this presentation is obtained by simplifying a Wirtinger presentation; the generators $x_1, \cdots, x_4$ are four of the original eight. Accordingly, we set $\mathfrak{a}\gamma x_i = t$, $i = 1, \cdots, 4$. The Alexander matrix $A = \| a_{ij} \|$ is given by

$$a_{ij} = \mathfrak{a}\gamma \frac{\partial}{\partial x_j} (x_i - [x_{i+3}, x_{i+2}^{-1}]x_{i+1}[x_{i+2}^{-1}, x_{i+3}]).$$

Hence,

$$\left.\begin{aligned} a_{ii} &= 1 \\ a_{i,i+1} &= -1 \\ a_{i,i+2} &= -t + 2 - t^{-1} \\ a_{i,i+3} &= t - 2 + t^{-1} \end{aligned}\right\} \quad \begin{aligned} &i = 1, \cdots, 4; \text{ indices} \\ &\text{are integers mod } 4. \end{aligned}$$

Any one of the four relations is a consequence of the other three and may be discarded. As a result, we may drop the 4th row of the matrix and obtain

$A \sim$		1	2	3	4
	1	1	-1	$-t + 2 - t^{-1}$	$t - 2 + t^{-1}$
	2	$t - 2 + t^{-1}$	1	-1	$-t + 2 - t^{-1}$
	3	$-t + 2 - t^{-1}$	$t - 2 + t^{-1}$	1	-1

The reader should check the operations in the following reduction of A to an equivalent matrix of simpler form.

$$A \sim \left\| \begin{matrix} 1 & -1 & -t+2-t^{-1} & 0 \\ t-2+t^{-1} & 1 & -1 & 0 \\ 0 & t-1+t^{-1} & 0 & 0 \end{matrix} \right\|$$

$$A \sim \left\| \begin{matrix} 1 & 0 & -t+2-t^{-1} & 0 \\ t-2+t^{-1} & 0 & -1 & 0 \\ 0 & t-1+t^{-1} & 0 & 0 \end{matrix} \right\|$$

$$A \sim \left\| \begin{matrix} 0 & 0 & -t+1-t^{-1} & 0 \\ t-3+t^{-1} & 0 & -1 & 0 \\ 0 & t-1+t^{-1} & 0 & 0 \end{matrix} \right\|$$

$$A \sim \left\| \begin{matrix} (t-3+t^{-1})(-t+1-t^{-1}) & 0 & 0 & 0 \\ 0 & 0 & -1 & 0 \\ 0 & t-1+t^{-1} & 0 & 0 \end{matrix} \right\|$$

$$A \sim \left\| \begin{matrix} (t-3+t^{-1})(t-1+t^{-1}) & 0 & 0 \\ 0 & t-1+t^{-1} & 0 \end{matrix} \right\|$$

Hence, the normalized polynomials are

$$\Delta_1 = (t^2 - 3t + 1)(t^2 - t + 1)^2,$$
$$\Delta_2 = t^2 - t + 1,$$
$$\Delta_k = 1 \quad \text{for} \quad k \geq 3.$$

So this knot is neither trivial, nor the clover-leaf, nor the figure-eight. Notice that the elementary ideals E_1 and E_2 are both principal ideals: E_1 generated by Δ_1 and E_2 by Δ_2.

(4.5) *Stevedore's knot* (Figure 50).

x and y are Wirtinger generators, and we set $\mathfrak{a}\gamma x = \mathfrak{a}\gamma y = t$. Using (3.7), we have immediately

$$\begin{aligned} \Delta_1 &= \mathfrak{a}\gamma \frac{\partial}{\partial x} [(xy^{-1})^{-2}y(xy^{-1})^2x - y(xy^{-1})^{-2}y(xy^{-1})^2] \\ &= -2 + 2t + t - (-2t + 2t^2) \\ &= -2t^2 + 5t - 2. \end{aligned}$$

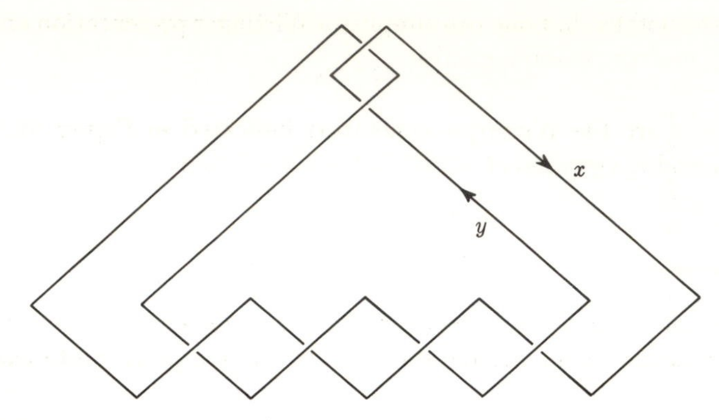

Figure 50. $\pi(R^3 - K) = | x,y : (xy^{-1})^{-2}y(xy^{-1})^2x = y(xy^{-1})^{-2}y(xy^{-1})^2 |$

The higher polynomials are, of course, all equal to 1. On normalizing, we obtain

$$\Delta_1 = 2t^2 - 5t + 2,$$
$$\Delta_k = 1 \quad \text{for} \quad k \geq 2.$$

The higher elementary ideals are also trivial.

$$E_1 = (2t^2 - 5t + 2),$$
$$E_k = (1) \quad \text{for} \quad k \geq 2.$$

(4.6) (Figure 51).

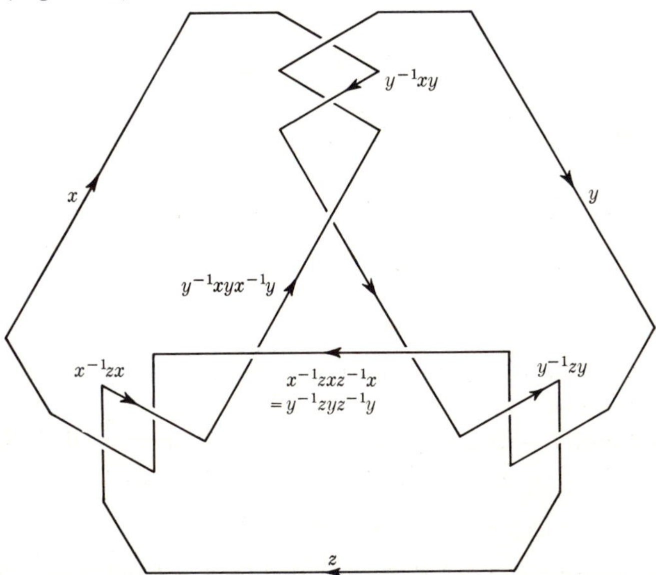

Figure 51

We omit the details; but one can simplify a Wirtinger presentation and obtain

$$\pi(R^3 - K) = | \; x,y,z \; : \; r_1,r_2 \; |$$

where x, y, z are the Wirtinger generators indicated in Figure 51, and the relators r_1 and r_2 correspond to the following relations.

$$r_1 \; : \; y^{-1}xyx^{-1}y = x^{-1}zx^{-1}zxz^{-1}x,$$

$$r_2 \; : \; x^{-1}zxz^{-1}x = y^{-1}zyz^{-1}y.$$

Under the mapping $\mathfrak{a}\gamma$ the generators x, y, z are each sent into t, and the Alexander matrix of the presentation can be written down quite easily. We get

$$a_{11} = \mathfrak{a}\gamma\left(\frac{\partial r_1}{\partial x}\right) = 3t^{-1} - 3$$

$$a_{12} = \mathfrak{a}\gamma\left(\frac{\partial r_1}{\partial y}\right) = -t^{-1} + 2$$

$$a_{13} = \mathfrak{a}\gamma\left(\frac{\partial r_1}{\partial z}\right) = -2t^{-1} + 1$$

$$a_{21} = \mathfrak{a}\gamma\left(\frac{\partial r_2}{\partial x}\right) = -t^{-1} + 2$$

$$a_{22} = \mathfrak{a}\gamma\left(\frac{\partial r_2}{\partial y}\right) = t^{-1} - 2$$

$$a_{23} = \mathfrak{a}\gamma\left(\frac{\partial r_2}{\partial z}\right) = 0.$$

Consequently,

$$A = \left\| \begin{matrix} 3t^{-1} - 3 & -t^{-1} + 2 & -2t^{-1} + 1 \\ -t^{-1} + 2 & t^{-1} - 2 & 0 \end{matrix} \right\|$$

$$\sim \left\| \begin{matrix} 3 - 3t & -1 + 2t & 0 \\ -1 + 2t & 1 - 2t & 0 \end{matrix} \right\|$$

$$\sim \left\| \begin{matrix} 2 - t & 0 & 0 \\ 0 & 1 - 2t & 0 \end{matrix} \right\|.$$

Since $2 - t$ and $1 - 2t$ are distinct irreducibles, their g.c.d. is 1. Hence,

$$\Delta_1 = (2 - t)(1 - 2t) = 2 - 5t + 2t^2,$$

$$\Delta_k = 1 \quad \text{for} \quad k \geq 2.$$

The second elementary ideal E_2 is generated by $2 - t$ and $2t - 1$. That this ideal is not the whole group ring JH of the abelianized group of the presentation may be seen by mapping JH homomorphically onto the integers J by setting $t \to -1, t^{-1} \to -1$. Under this homomorphism the ideal E_2 is mapped onto the integral ideal generated by 3 since

$$2 - t \quad \to 3,$$
$$1 - 2t \quad \to 3.$$

We conclude

$$E_1 = (2t^2 - 5t + 2),$$
$$E_2 = (2 - t, 1 - 2t) \text{ is not a principal ideal,}$$
$$E_k = (1) \quad \text{for} \quad k \geq 3.$$

Comparison with the preceding example shows that the two knots exhibited in Figures 50 and 51 have *the same knot polynomials but distinct elementary ideals*. These examples verify the contention made in Chapter VII and in the introduction to this chapter that the elementary ideals are stronger invariants than the polynomials.

(4.7) (Figures 52 and 53).

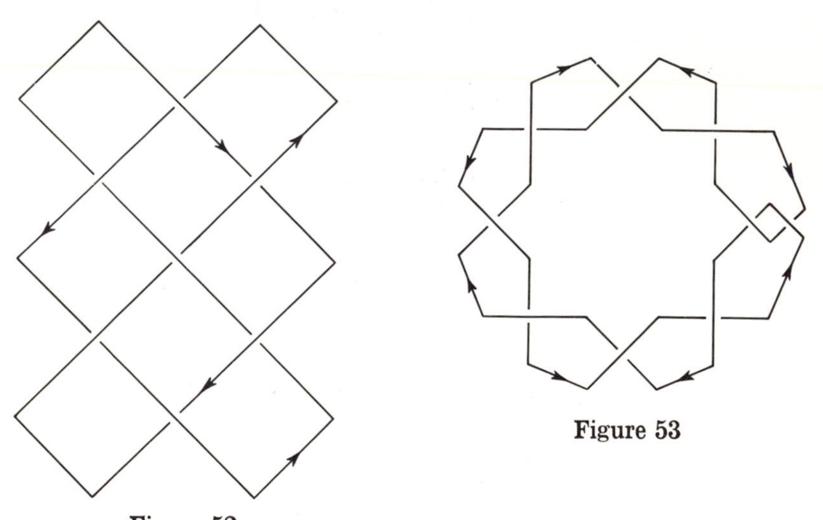

Figure 52

Figure 53

The Alexander matrix of each of these knot types is equivalent to the matrix

$$\| \, 4t^2 - 7t + 4 \qquad 0 \, \|.$$

Thus, *the methods developed in the last two chapters fail to distinguish them*. Their groups can, however, be shown to be nonisomorphic by other methods.[3]

[3] By the linking invariant of the second cyclic branched covering; cf. H. Seifert, "Die Verschlingungsinvarianten der zyklischen Knotenüberlagerungen," *Hamb. Abh.* 11 (1935) pp. 84–101.

(4.8) *Granny knot and Square knot* (Figures 54 and 55).

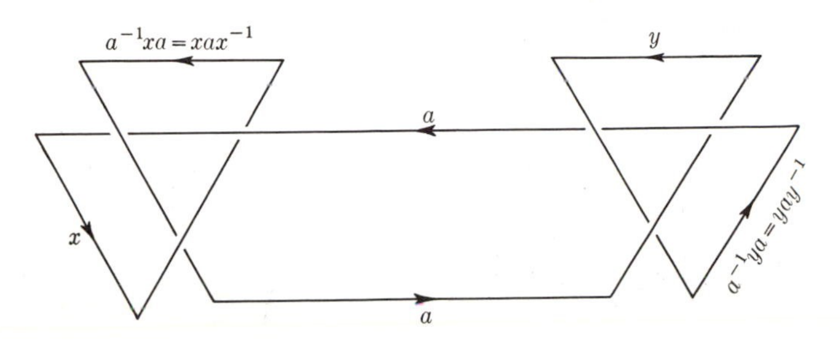

Figure 54

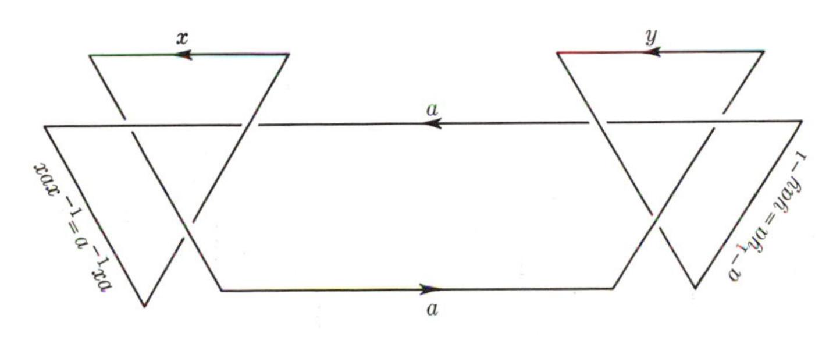

Figure 55

It can be shown by more advanced techniques that the knots shown in Figure 54 and Figure 55 represent distinct knot types.[4] However, the methods of this book fail to distinguish them from the very outset. Not only do they have equivalent Alexander matrices, but *they even possess isomorphic groups*. For each, we have

$$\pi(R^3 - K) = |\ x,y,a\ :\ a^{-1}xa = xax^{-1},\ a^{-1}ya = yay^{-1}\ |.$$

EXERCISES

1. For each of the five knots in Exercise 1 of Chapter VI, find an Alexander matrix with one row and two columns. Compute the elementary ideals and knot polynomials.

---

[4] R. H. Fox, "On the Complementary Domains of a Certain Pair of Inequivalent Knots," *Ned. Akademie Wetensch., Indag. Math.* Vol. 14 (1952), pp. 37–40; H. Seifert, "Verschlingungsinvarianten," *S. B. Preuss. Akad. Wiss. Berlin* Vol. 26 (1933), pp. 811–823.

2. Compute the elementary ideals and knot polynomials for each of the following four knots.

(a) True lover's knot.
(b) False lover's knot.
(c) Chinese button knot.
(d) Bowline knot.

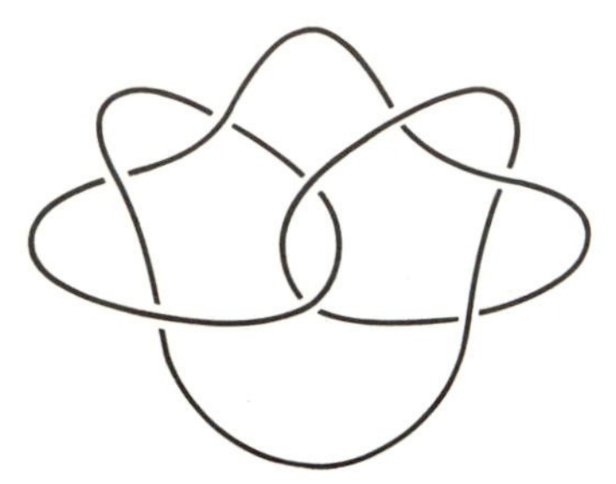

Figure 56a

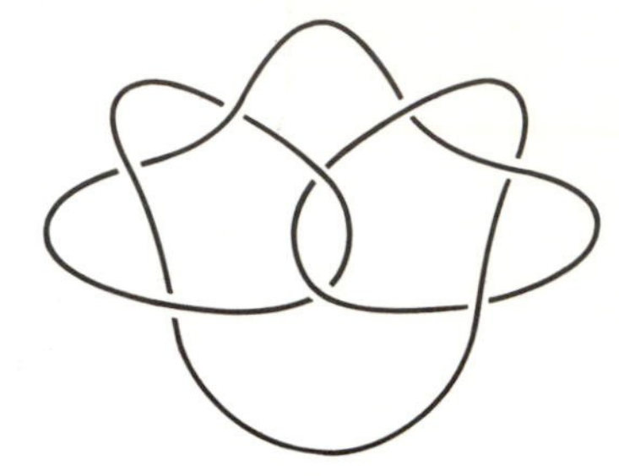

Figure 56b

Figure 56c

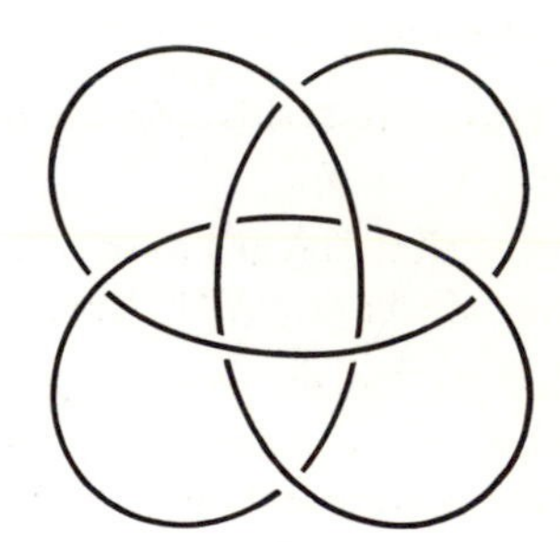

Figure 56d

3. Using the presentation of the group of the torus knot $K_{p,q}$ given in Chapter VI, Exercise 3, show that the Alexander polynomial is

$$\Delta(t) = \frac{(t^{pq} - 1)(t - 1)}{(t^p - 1)(t^q - 1)},$$

and $E_2 = 1$.

4. Prove that the degree of the normalized Alexander polynomial of a knot is not greater than the number of crossings of any of its diagrams.

5. Show that the Alexander matrix class of the knot in Figure 57 is the same as that of a trivial knot.

6. Prove that the group of the figure-eight cannot be mapped homomorphically upon the group of the overhand knot.

7. If we tie two knots on the same piece of string, the result is called a composite knot. Prove that the Alexander polynomial of a composite knot is the product of the polynomials of the constituent knots.

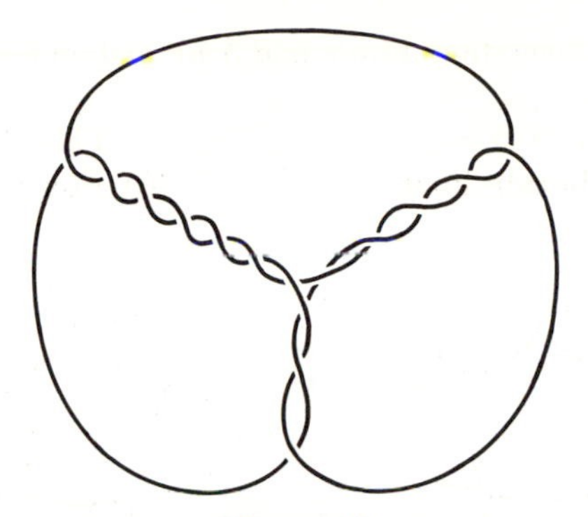

Figure 57

8. Let k and n be positive integers and let us try to assign to each overpass x_j of a regular knot projection an integer λ_j in such a way that corresponding to each crossing $x_j \longrightarrow \underset{x_k}{\downarrow} \longrightarrow x_{j+1}$ the relation $k(\lambda_k - \lambda_j) \equiv \lambda_k - \lambda_{j+1} \pmod{n}$ holds. Prove that this can be done nontrivially if and only if $\Delta(k) \equiv 0 \pmod{n}$.

9. Prove that the group of a knot can be mapped homomorphically upon the group of Chapter VII, Exercise 12, if and only if $\Delta(k) \equiv 0 \pmod{p}$.

10. Let $f(t)$ be an integral polynomial and let us try to assign to each overpass x_j of a regular knot projection an integral polynomial $\Lambda_j(t)$ in such a way that corresponding to each crossing as shown in Exercise 8 the relation $t(\Lambda_k(t) - \Lambda_j(t)) \equiv \Lambda_k(t) - \Lambda_{j+1}(t) \pmod{f(t)}$ holds. Prove that this can be done nontrivially if and only if $f(t)$ divides $\Delta(t)$.

CHAPTER IX

Characteristic Properties of the Knot Polynomials

Introduction. A survey of the knot polynomials $\Delta_k(t)$ computed at the end of the preceding chapter shows that, for each of them, $\Delta_k(1) = \pm 1$. A proof that this equation holds for all knot polynomials is the objective of the first section of the present chapter. The survey also substantiates the assertion that all knot polynomials are *reciprocal polynomials*, i.e., for every knot polynomial $\Delta_k(t)$, there exists an integer n such that $\Delta_k(t) = t^n \Delta_k(t^{-1})$. Thus, if $\Delta_k(t) = c_n t^n + c_{n-1} t^{n-1} + \cdots + c_0$, the coefficients exhibit the symmetry $c_i = c_{n-i}$, $i = 0, \cdots, n$. As was pointed out in Section 3 of Chapter VIII, this property is essential to our conclusion that knots of the same type possess identical polynomials. It is therefore important to close this gap in the theory. The proof that knot polynomials are reciprocal polynomials will be effected in Sections 2 and 3 by introducing the notion of dual group presentations, the crucial examples of which are the over and under presentations of knot groups defined in Chapter VI. It should be emphasized that our arguments apply only to tame knots, and throughout this chapter "knot" always means "tame knot."

It is known[1] that the two properties

$\Delta_1(1) = \pm 1$,
$\Delta_1(t)$ is a reciprocal polynomial,

characterize the 1st polynomial or Alexander polynomial $\Delta_1(t)$ of a knot; in other words any L-polynomial that has these two properties is the 1st polynomial of some knot.

1. Operation of the trivializer. An element a of the group ring JH of an infinite cyclic group H has a representation as an L-polynomial

$$a = a(t) = \sum_{-\infty}^{\infty} a_n t^n,$$

where all but a finite number of the integers a_n are equal to zero. The image of a

[1] H. Seifert, "Über das Geschlect von Knoten," *Math. Ann.* Vol. 110 (1934), pp. 571–592; G. Torres and R. H. Fox, "Dual Presentations of the Group of a Knot," *Ann. of Math.* Vol. 59 (1954), pp. 211–218.

under the trivializing homomorphism $\mathfrak{t} : JH \to J$ is obtained by setting $t = 1$ (cf. Section 1, Chapter VII). Thus, we write $\mathfrak{t}a = a(1)$. An invariant of knot type, simpler than the knot polynomials $\Delta_k(t)$, is the sequence of integers $| \Delta_k(1) |$, $k = 1, 2, \cdots$. Although each knot polynomial is specified only to within a unit factor $\pm t^n$, the absolute value $| \Delta_k(1) |$ is uniquely determined by the isomorphism class of the knot group. This invariant, however, is useless as a tool for distinguishing knot types. In this section we shall prove the interesting theorem that, for any knot group,

$$(1.1) \qquad | \Delta_k(1) | = 1, \qquad k = 1, 2, \cdots.$$

An equivalent result is

(1.2) *For any finite presentation* $(\mathbf{x} : \mathbf{r})$ *of a knot group and integer* $k \geq 1$, *the image of the elementary ideal* E_k *of* $(\mathbf{x} : \mathbf{r})$ *under the trivializer* $\mathfrak{t}$ *is the entire ring of integers,* i.e., $\mathfrak{t}E_k = J$, $k = 1, 2 \cdots$.

It is easy to show that (1.1) and (1.2) are equivalent. Observe, first of all, that since $\Delta_{k+1} \mid \Delta_k$ (cf. (3.3), Chapter VIII), (1.1) is equivalent to the statement that $| \Delta_1(1) | = 1$. Similarly, the elementary ideals form an ascending chain, and so (1.2) is equivalent to the equation $\mathfrak{t}E_1 = J$. We have shown that E_1 is a principal ideal generated by Δ_1. It follows that $\mathfrak{t}E_1$ is generated by $\mathfrak{t}\Delta_1 = \Delta_1(1)$. Thus, if $| \Delta_1(1) | = 1$, then $\mathfrak{t}E_1 = J$. Conversely, since the generator of an ideal in an integral domain is unique to within units, if $\mathfrak{t}E_1 = J$, then $| \Delta_1(1) | = 1$.

We now prove (1.2). Let $(\mathbf{x} : \mathbf{r}) = (x_1, \cdots, x_n : r_1, \cdots, r_n)$ be a finite presentation of a knot group and A its Alexander matrix. As a result of (4.6) of Chapter IV, the abelianized group of the knot group can be presented by $(\mathbf{x} : \mathbf{r}, [x_i,x_j], i, j = 1, \cdots, n)$. Denote the Alexander matrix of this presentation by A'. Since the abelianized group of any knot group is infinite cyclic, $(\mathbf{x} : \mathbf{r}, [x_i,x_j], i, j = 1, \cdots, n)$ is of the same presentation type as $(x :)$. The elementary ideals of the latter are $E_0 = (0)$, $E_1 = E_2 = \cdots = (1)$. It follows from the fundamental invariance theorem for elementary ideals (cf. (4.5), Chapter VII) that

$$E_k(A') = \begin{cases} (0), & k = 0, \\ (1), & k \geq 1. \end{cases}$$

The ideal (1), generated by the identity 1, is, of course, the entire ring. We next observe that the image of any Alexander matrix under the trivializer is identical with the image of the original matrix of free derivatives under the trivializer. Furthermore,

$$\frac{\partial}{\partial x_k} [x_i,x_j] = \delta_{ik}(1 - x_ix_jx_i^{-1}) + \delta_{jk}(x_i - x_ix_jx_i^{-1}x_j^{-1}).$$

And so,

$$\mathfrak{t}\frac{\partial}{\partial x_k} [x_i,x_j] = 0, \qquad i, j, k = 1, \cdots, n.$$

Hence,

$$\mathrm{t}A' = \begin{pmatrix} \left\| \mathrm{t}\dfrac{\partial r_i}{\partial x_j} \right\| \\ 0 \end{pmatrix} = \begin{pmatrix} \mathrm{t}A \\ 0 \end{pmatrix} \sim \mathrm{t}A.$$

Using the results of this paragraph and (4.2) and (4.3) of Chapter VII, we obtain

$$\begin{aligned} \mathrm{t}E_k(A) &= E_k(\mathrm{t}A) = E_k(\mathrm{t}A') = \mathrm{t}E_k(A') \\ &= \begin{cases} (0), & \text{if } k = 0, \\ J, & \text{if } k \geq 1, \end{cases} \end{aligned}$$

and the proof of (1.2) is complete.

2. Conjugation. The immediate objective of this and the next section is the theorem:

(2.1) *For any knot polynomial $\Delta_k(t)$, there exists an even integer n such that*

$$\Delta_k(t) = t^n \Delta_k\left(\frac{1}{t}\right).$$

Notice that if (2.1) holds for a polynomial $\Delta_k(t)$, it also holds for any unit multiple of $\Delta_k(t)$.

The *degree* of an arbitrary L-polynomial $a(t) = \Sigma_{-\infty}^{\infty} a_m t^m$ ($a_m = 0$ for all but a finite number of values) is the difference between the largest and smallest values of m for which $a_m \neq 0$. Since this number is unaffected by multiplication by a unit factor $\pm t^k$ or by the change of variable $s = t^{-1}$, the degree of a knot polynomial is a well-defined invariant of knot type. If the polynomial $\Delta_k(t)$ is chosen in normalized form (no negative powers of t and a positive constant term), then the integer n which appears in the statement of (2.1) above is obviously the degree of $\Delta_k(t)$. Thus, in addition to stating that knot polynomials are reciprocal polynomials, (2.1) implies that

(2.2) *Every knot polynomial is of even degree.*

The mapping $(\ )^{-1}\colon G \to G$ which assigns to every element g of an arbitrary group G its inverse g^{-1}, is one-one and onto but not an isomorphism (unless G is abelian). Since it is product-reversing instead of preserving, i.e.,

$$(gh)^{-1} = h^{-1}g^{-1}, \qquad g, h \in G,$$

it is called an *anti-isomorphism*. An important fact, albeit trivially verifiable, is that $(\ )^{-1}$ is consistent with homomorphisms: Given any homomorphism $\phi\colon G \to H$, the mapping diagram is consistent. The unique linear extension

$$\begin{array}{ccc} G & \xrightarrow{(\ )^{-1}} & G \\ \phi\downarrow & & \downarrow\phi \\ H & \xrightarrow{(\ )^{-1}} & H \end{array}$$

of ($)^{-1}$ to the group ring JG (cf. (1.2), Chapter VII) will be called *conjugation* and denoted by a bar. Thus,

$$\overline{\sum a_i g_i} = \sum a_i (g_i)^{-1}, \qquad a_i \in J,\ g_i \in G.$$

Using the theory of dual presentations developed in the next section, we shall prove the important theorem

(2.3) *The elementary ideals E_k of any finite presentation of a knot group are invariant under conjugation*, i.e., $\overline{E}_k = E_k$, $k = 0, 1, 2, \cdots$.

Theorem (2.1) is a corollary of (2.3) and (1.1). The proof is as follows: Denote by (Δ_k) the principal ideal generated by the knot polynomial Δ_k. We recall that (Δ_k) is the smallest principal ideal containing E_k. Since $E_k \subset (\Delta_k)$,

$$E_k = \overline{E}_k \subset \overline{(\Delta_k)}.$$

In the group ring of an abelian group, conjugation is a ring isomorphism. Hence, $\overline{(\Delta_k)}$ is a principal ideal, and $(\overline{\Delta}_k) = \overline{(\Delta_k)}$. Since (Δ_k) is minimal,

$$(\Delta_k) \subset \overline{(\Delta_k)} \subset \overline{\overline{(\Delta_k)}} = (\Delta_k).$$

We conclude that

$$(\Delta_k) = (\overline{\Delta}_k).$$

Generators of a principal ideal in an integral domain are unique to within units; hence,

$$\Delta_k(t) = \epsilon t^n \Delta_k\left(\frac{1}{t}\right),$$

where $\epsilon = \pm 1$. (Of course, $\Delta_k(t) = \Delta_k$ and $\Delta_k\left(\frac{1}{t}\right) = \overline{\Delta}_k$.) For $k = 0$, both sides of the equation are zero, and the value of ϵ doesn't much matter. For $k > 0$, we know from (1.1) that $\Delta_k(1) \neq 0$. Hence, substituting $t = 1$ gives immediately $\epsilon = 1$. Writing $\Delta_k(t) = c_0 + c_1 t + \cdots + c_n t^n$, we have $c_i = c_{n-i}$, $i = 0, \cdots, n$. If n were odd, we would have by (1.1)

$$|\ \Delta_k(1)\ | = 1 = 2\ |\ c_0 + \cdots + c_{(n-1)/2}\ |,$$

which is impossible. Hence, n is even, and the proof of (2.1) from (2.3) and (1.1) is complete.

3. Dual presentations. The definition of dual presentations is conveniently expressed using the terminology of congruences. If $f\colon R \rightarrow R'$ is any ring homomorphism and $a_1, a_2 \in R$, we write $a_1 \equiv a_2 \pmod f$, translated a_1 *is congruent to* a_2 *modulo* f, whenever $fa_1 = fa_2$. (The expression appears most commonly in consideration of the homomorphism of the integers J onto the ring J_n of residue classes.) Two finite group presentations $(\mathbf{x} : \mathbf{r}) = (x_1, \cdots, x_n : r_1, \cdots, r_n)$ and $(\mathbf{y} : \mathbf{s}) = (y_1, \cdots, y_n : s_1, \cdots, s_n)$ constitute a pair of *dual presentations* if there exists a presentation equivalence $\theta\colon (\mathbf{x} : \mathbf{r}) \to (\mathbf{y} : \mathbf{s})$ such that

$$(i)\ \theta x_i \equiv y_i^{-1} \qquad (\text{mod } \mathfrak{a}\gamma), \quad i = 1, \cdots, n,$$

$$(ii)\ \theta\left(\frac{\partial r_i}{\partial x_j}(x_j - 1)\right) \equiv \overline{\frac{\partial s_j}{\partial y_i}(y_i - 1)} \qquad (\text{mod } \mathfrak{a}\gamma), \quad i, j = 1, \cdots, n.$$

The homomorphism γ is the extension to the group ring of the canonical homomorphism of the free group generated by $y_1, \cdots, y_n$ onto the factor group $|\, \mathbf{y} : \mathbf{s} \,|$, and $\mathfrak{a}$ is the abelianizer.

Dual presentations are *ipso facto* of the same type, and it therefore makes sense to speak of dual presentations of a given group. It is our contention that the group of any knot has a pair of dual presentations. Specifically, we shall prove that the over and under presentations (1.1) and (1.2), Chapter VI, are mutually dual. We assume that K is a polygonal knot in regular position situated as described in Section 1, Chapter VI, and that overpasses, underpasses, orientations, basepoints, generators, etc. have been selected as there described. The notation will be the same. The presentations (1.1) and (1.2) are abbreviated $(\mathbf{x} : \mathbf{r})$ and $(\mathbf{y} : \mathbf{s})$ respectively, and the canonical homomorphisms of $F(\mathbf{x})$ and $F(\mathbf{y})$ onto the factor groups $|\, \mathbf{x} : \mathbf{r} \,|$ and $|\, \mathbf{y} : \mathbf{s} \,|$ are both denoted by γ. Let α be the equivalence class of a path in $R^3 - K$ with initial point p_0' and terminal point p_0. The mapping η defined by

$$\eta\beta = \alpha \cdot \beta \cdot \alpha^{-1}$$

for all $\beta \in \pi(R^3 - K, p_0)$ is an isomorphism of $\pi(R^3 - K, p_0)$ onto $\pi(R^3 - K, p_0')$, cf. (3.1), Chapter II. This isomorphism induces a natural isomorphism θ_* of $|\, \mathbf{x} : \mathbf{r} \,|$ onto $|\, \mathbf{y} : \mathbf{s} \,|$ which is realizable by a presentation equivalence $\theta\colon (\mathbf{x} : \mathbf{r}) \to (\mathbf{y} : \mathbf{s})$ (cf. (2.3) and (2.4), Chapter IV). All of these actually very simple ideas are summed up in the following completely consistent diagram.

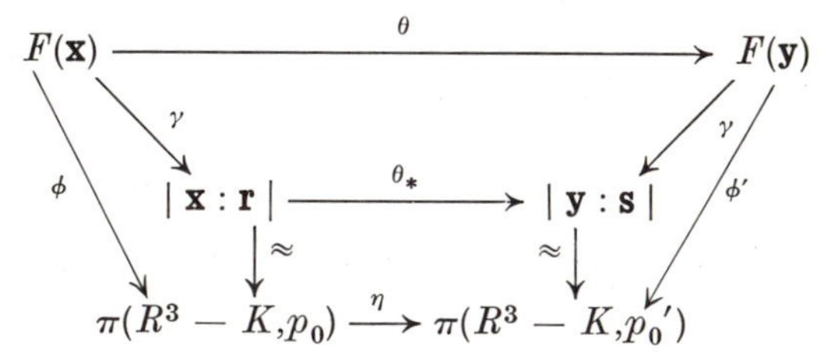

Consider an arbitrary underpass B_k adjacent to an overpass A_j. Where β is the equivalence class of the path shown in Figure 58 below, it is clear that

$$\eta\phi x_j = (\alpha \cdot \beta^{-1}) \cdot (\beta \cdot \phi x_j \cdot \beta^{-1}) \cdot (\beta \cdot \alpha^{-1}),$$

and that

$$\beta \cdot \phi x_j \cdot \beta^{-1} = (\phi' y_k)^{-1}.$$

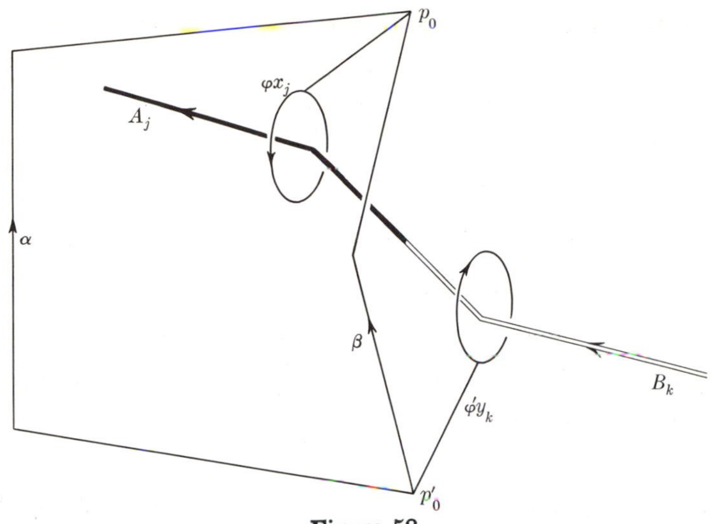

Figure 58

Hence,

$$\eta\phi x_j = (\alpha \cdot \beta^{-1}) \cdot (\phi' y_k^{-1}) \cdot (\beta \,.\, \alpha^{-1}),$$

and so

$$\theta_* \gamma x_j = \sigma \cdot (\gamma y_k^{-1}) \cdot \sigma^{-1},$$

where σ is the image of $(\alpha \cdot \beta^{-1})$ under the isomorphism $\pi(R^3 - K, p_0') \approx |\, \mathbf{y} : \mathbf{s} \,|$. We denote the abelianizer on $|\, \mathbf{x} : \mathbf{r} \,|$ and on $|\, \mathbf{y} : \mathbf{s} \,|$ by the same letter $\mathfrak{a}$. Finally, therefore,

$$\mathfrak{a}\gamma\theta x_j = \mathfrak{a}\theta_* \gamma x_j = \mathfrak{a}[\sigma \cdot (\gamma y_k^{-1}) \cdot \sigma^{-1}] = \mathfrak{a}\gamma y_k^{-1}.$$

This coupled with the fact (cf. (1.1), Chapter VIII) that

$$(3.1) \quad x_i \equiv x_j \qquad (\text{mod } \mathfrak{a}\gamma), \qquad i, j = 1, \cdots, n,$$

implies

$$\theta x_k \equiv y_k^{-1} \qquad (\text{mod } \mathfrak{a}\gamma),$$

and Condition (i) of the definition of dual presentations is established.

It is a corollary of the preceding two equations that

$$(3.2) \quad y_i \equiv y_j \qquad (\text{mod } \mathfrak{a}\gamma), \qquad i, j = 1, \cdots, n.$$

Moreover (cf. (1.1) and (1.2), Chapter VIII), the infinite cyclic abelianized group of $|\, \mathbf{y} : \mathbf{s} \,|$ is generated by the single element $s = \mathfrak{a}\gamma y_i$, $i = 1, \cdots, n$.

In order to establish (*ii*), we shall need the following lemma:

(3.3) *Let a and b be any two simple paths in $R^2 - \mathscr{P}B$ and $R^2 - \mathscr{P}A$, respectively, having the same initial and terminal points* (i.e., $a(0) = b(0)$ and $a(\| a \|) = b(\| b \|)$). *Then,*

$$\theta a^{\#} \equiv (b^{\flat})^{-1} \pmod{\mathfrak{a}\gamma}.$$

Proof. As a result of (3.1) and (3.2), we have

$$a^{\#} \equiv x_1^{\,l} \pmod{\mathfrak{a}\gamma}, \quad \text{for some integer } l,$$

$$b^{\flat} \equiv y_1^{\,m} \pmod{\mathfrak{a}\gamma}, \quad \text{for some integer } m.$$

Next, choose a simple path c in $R^2 - \mathscr{P}A$ with the same initial and terminal points as a and b such that

$$c^{\flat} \equiv y_1^{\,l} \pmod{\mathfrak{a}\gamma}.$$

There are several ways to pick c. For example, one may simply follow along a as closely as possible skirting around every projected overpass encountered (cf. Figure 59). Suppose, for example, that the path a crosses under the

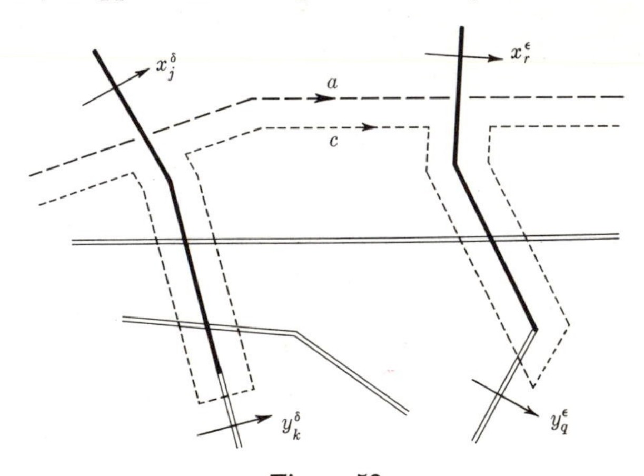

Figure 59

overpass A_j so that the resulting contribution to $a^{\#}$ is $x_j^{\,\delta}$. Then, as is illustrated in Figure 59, in skirting around $\mathscr{P}A_j$ the contribution to c obtained by crossing underpasses is

$$(y_{i_1}^{\epsilon_1} \cdots y_{i_p}^{\epsilon_p}) y_k^{\,\delta} (y_{i_p}^{-\epsilon_p} \cdots y_{i_1}^{-\epsilon_1}).$$

Thus the exponent sum of $a^{\#}$ and $c^{\flat}$ must be equal. Since cb^{-1} is a closed path which cuts no projected overpasses, we have

$$\phi'(c^{\flat}(b^{\flat})^{-1}) = 1.$$

Hence,

$$c^\flat \equiv b^\flat \qquad (\text{mod } \mathfrak{a}\gamma),$$

and so

$$y_1{}^l \equiv y_1{}^m \qquad (\text{mod } \mathfrak{a}\gamma).$$

Since $\mathfrak{a}\gamma y_1$ generates an infinite cyclic group, we conclude that $l = m$. Finally, using Condition (i), we obtain

$$\theta a^\# \equiv \theta x_1{}^m \equiv y_1{}^{-m} \equiv (b^\flat)^{-1} \qquad (\text{mod } \mathfrak{a}\gamma),$$

and the proof is complete. Notice, incidentally, that (3.3) includes (i) as a special case.

The presentations $(\mathbf{x} : \mathbf{r})$ and $(\mathbf{y} : \mathbf{s})$ are unaffected by the size and shape of the regions $V_1, \cdots, V_n$ and $U_1, \cdots, U_n$. Consequently, we shall assume that the points of each one remain close to the particular projected underpass or overpass covered. Consider an arbitrary pair of integers $i, j = 1, \cdots, n$. We have (cf. (1.1) and (1.2), Chapter VI) $r_i = c_i^\# \cdot v_i^\# \cdot (c_i^\#)^{-1}$ and $s_j = d_j^\flat \cdot u_j^\flat \cdot (d_j^\flat)^{-1}$. Notice that

$$\frac{\partial r_i}{\partial x_j} \equiv \frac{\partial c_i^\#}{\partial x_j} + c_i^\# \frac{\partial v_i^\#}{\partial x_j} - \frac{\partial c_i^\#}{\partial x_j} \equiv c_i^\# \frac{\partial v_i^\#}{\partial x_j} \qquad (\text{mod } \gamma),$$

and similarly

$$\frac{\partial s_j}{\partial y_i} \equiv d_j^\flat \frac{\partial u_j^\flat}{\partial y_i} \qquad (\text{mod } \gamma).$$

Thus, in checking (ii), we need only consider occurrences of x_j in $v_i^\#$ and y_i in $u_j^\flat$. We shall say that the overpass A_j is *adjacent* to the underpass B_i if they have an endpoint in common, i.e., if they occur consecutively along K. The different cases may be classified as follows:

CASE (1) *The overpass A_j neither crosses over nor is adjacent to the underpass B_i.* In this case, $v_i^\#$ does not contain x_j and $u_j^\flat$ does not contain y_i. Hence,

$$\frac{\partial r_i}{\partial x_j} \equiv \frac{\partial s_j}{\partial y_i} \equiv 0 \qquad (\text{mod } \gamma).$$

CASE (2) *The overpass A_j crosses over B_i at least once, but is not adjacent to B_i.* We include the possibility that A_j crosses B_i several times. However, each intersection of v_i with $\mathscr{P}A_j$ contributes a monomial term to $\frac{\partial r_i}{\partial x_j}$, and $\frac{\partial r_i}{\partial x_j}$ is just the sum of these contributions. Similarly, $\frac{\partial s_j}{\partial y_i}$ is the sum of the monomials contributed by the intersections of u_j and $\mathscr{P}B_i$. Thus we may study one crossing at a time. The situation at a single crossing of B_i by A_j is shown in Figure 60.

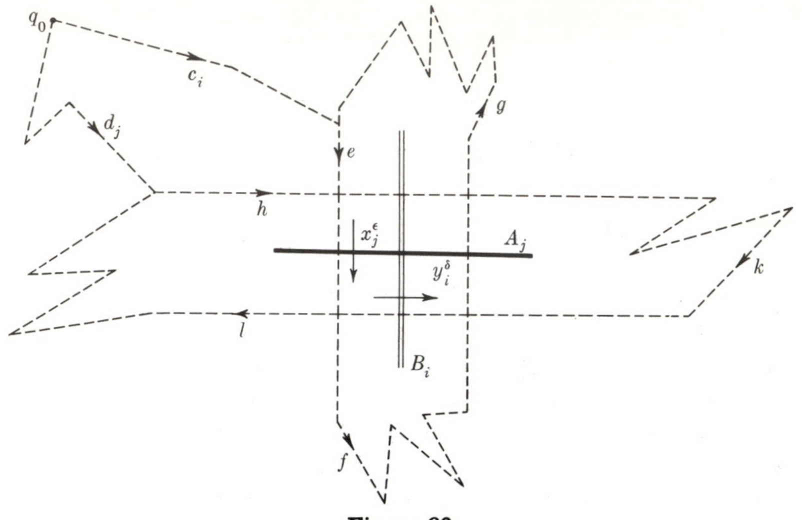

Figure 60

Then,

$$r_i = c_i^{\#}(e^{\#}x_j^{\epsilon}f^{\#}x_j^{-\epsilon}g^{\#})(c_i^{\#})^{-1},$$
$$s_j = d_j^{\flat}(h^{\flat}y_i^{\delta}k^{\flat}y_i^{-\delta}l^{\flat})(d_j^{\flat})^{-1}.$$

Hence,

$$\frac{\partial r_i}{\partial x_j} = (c_i^{\#}e^{\#} - c_i^{\#}e^{\#}x_j^{\epsilon}f^{\#}x_j^{-\epsilon})\,\frac{x_j^{\epsilon} - 1}{x_j - 1} + \cdots,$$
$$\frac{\partial s_j}{\partial y_i} = (d_j^{\flat}h^{\flat} - d_j^{\flat}h^{\flat}y_i^{\delta}k^{\flat}y_i^{-\delta})\,\frac{y_i^{\delta} - 1}{y_i - 1} + \cdots.$$

Therefore,

$$\frac{\partial r_i}{\partial x_j}(x_j - 1) \equiv c_i^{\#}e^{\#}(1 - f^{\#})(x_j^{\epsilon} - 1) + \cdots \qquad (\text{mod } \mathfrak{a}\gamma),$$
$$\frac{\partial s_j}{\partial y_i}(y_i - 1) \equiv d_j^{\flat}h^{\flat}(1 - k^{\flat})(y_i^{\delta} - 1) + \cdots \qquad (\text{mod } \mathfrak{a}\gamma).$$

By the lemma (3.3),

$$\theta c_i^{\#}e^{\#} \equiv (d_j^{\flat}h^{\flat})^{-1} \qquad (\text{mod } \mathfrak{a}\gamma),$$
$$\theta f^{\#} \equiv y_i^{-\delta} \qquad (\text{mod } \mathfrak{a}\gamma),$$
$$\theta x_j^{\epsilon} \equiv (k^{\flat})^{-1} \qquad (\text{mod } \mathfrak{a}\gamma).$$

Hence,

$$\theta\left(\frac{\partial r_i}{\partial x_j}(x_j - 1)\right) \equiv \overline{d_j^{\flat}h^{\flat}(1 - y_i^{\delta})(k^{\flat} - 1)} + \cdots \qquad (\text{mod } \mathfrak{a}\gamma),$$

and it follows that

$$\theta\left(\frac{\partial r_i}{\partial x_j}(x_j - 1)\right) \equiv \overline{\frac{\partial s_j}{\partial y_i}(y_i - 1)} \qquad (\text{mod } \mathfrak{a}\gamma).$$

CASE (3) *The overpass A_j is adjacent to B_i.* The situation is illustrated in Figure 61.

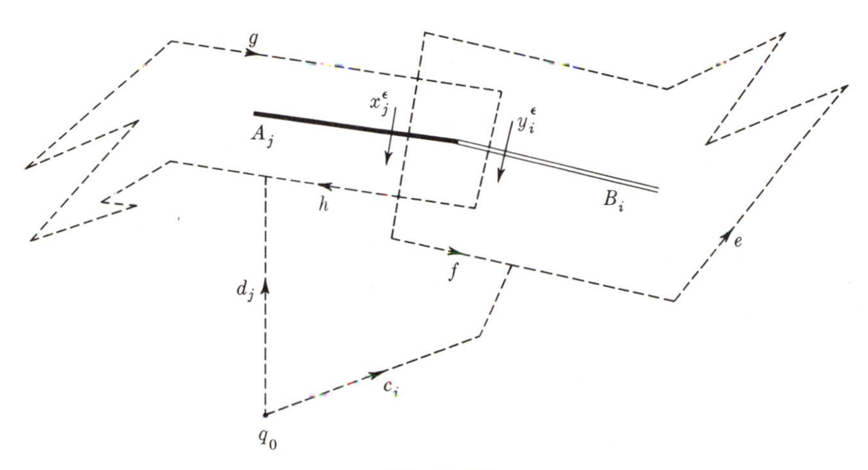

Figure 61

We have

$$r_i = c_i^\#(e^\# x_j^\epsilon f^\#)(c_i^\#)^{-1},$$
$$s_j = d_j^\flat(g^\flat y_i^\epsilon h^\flat)(d_j^\flat)^{-1}.$$

Hence,

$$\frac{\partial r_i}{\partial x_j}(x_j - 1) = c_i^\# e^\#(x_j^\epsilon - 1) + \cdots,$$
$$\frac{\partial s_j}{\partial y_i}(y_i - 1) = d_j^\flat g^\flat(y_i^\epsilon - 1) + \cdots.$$

By the Lemma (3.3),

$$\theta(c_i^\# e^\#) \equiv (d_j^\flat g^\flat)^{-1} \qquad (\text{mod } \mathfrak{a}\gamma),$$
$$\theta x_j^\epsilon \equiv y_i^{-\epsilon} \qquad (\text{mod } \mathfrak{a}\gamma),$$

and so

$$\theta\left(\frac{\partial r_i}{\partial x_j}(x_j - 1)\right) \equiv \overline{(d_j^\flat g^\flat)(y_i^\epsilon - 1)} + \cdots \qquad (\text{mod } \mathfrak{a}\gamma).$$

We have already observed in Case (2) that the contribution from the crossings of A_j over B_i yield terms that cancel in pairs (mod $\mathfrak{a}\gamma$). It follows that

$$\theta\left(\frac{\partial r_i}{\partial x_j}(x_j - 1)\right) \equiv \overline{\frac{\partial s_j}{\partial y_i}(y_i - 1)} \qquad (\text{mod } \mathfrak{a}\gamma),$$

and the proof that $(\mathbf{x} : \mathbf{r})$ and $(\mathbf{y} : \mathbf{s})$ are dual presentations is complete. We have proved

(3.4) *Corresponding over and under presentations of the group of a knot are dual presentations.*

The invariance under conjugation of the elementary ideals of a knot group, i.e., (2.3), is an easy corollary of Theorem (3.4). In view of (3.1) and (3.2) and Condition (i), we have the stronger result

$$\theta\left(\frac{\partial r_i}{\partial x_j}\right) \equiv \overline{\frac{\partial s_j}{\partial y_i}} \qquad (\text{mod } \mathfrak{a}\gamma) \qquad i, j = 1, \cdots, n.$$

That is, if A and B are the Alexander matrices of $(\mathbf{x} : \mathbf{r})$ and $(\mathbf{y} : \mathbf{s})$, respectively, then

(3.5) $\theta_{**}A = \overline{B^\dagger}$.

(The † indicates the transposed matrix.) The transpose of a square matrix obviously has the same elementary ideals as the original. Let us denote the kth elementary ideals of $(\mathbf{x} : \mathbf{r})$ and $(\mathbf{y} : \mathbf{s})$ by $E_k^{(1)}$ and $E_k^{(2)}$, respectively. The kth ideal of an arbitrary matrix M we denote by $E_k(M)$. Thus $E_k^{(1)} = E_k(A)$ and $E_k^{(2)} = E_k(B)$. Finally, therefore (cf. Chapter VII, (4.3) and (4.5)),

$$\begin{aligned} E_k^{(2)} = \theta_{**}E_k^{(1)} &= E_k(\overline{B}^\dagger) \\ &= E_k(\overline{B}) = \overline{E_k(B)} \\ &= \overline{E_k^{(2)}}. \end{aligned}$$

Hence Theorem (2.3) of this chapter is proved.

EXERCISES

1. Show that there exists an automorphism of the group G of the clover-leaf knot that induces conjugation on the group ring of the abelianized group G/G'.

2. Use the result of Exercise 1 to prove directly (i.e., without using dual presentations) that the elementary ideals of the group of the clover-leaf knot are invariant under conjugation.

3. Prove directly (i.e., without using dual presentations) that the elementary ideals of any invertible knot are invariant under conjugation.

4. Show that the Alexander polynomial $\Delta(t)$ of any knot can be written in the form

$$\Delta(t) = t^h + c_1 t^{h-1}(1-t)^2 + c_2 t^{h-2}(1-t)^4 + \cdots + c_h(1-t)^{2h},$$

and that, conversely, given any set of integers $c_1, \cdots, c_h$ there is a knot whose Alexander polynomial is

$$t^h + \sum_{i=1}^{h} c_i t^{h-i}(1-t)^{2i}.$$

5. Prove that there is no knot whose group is

$$| \, x, y \; : \; xyx^{-1}yx = yx^{-1}yxy \, |.$$

6. Prove that $\Delta(-1)$ is always an odd integer.

7. If the Alexander polynomial $\Delta(t)$ of a knot is of degree $2h$ and ϵ is a complex number on the unit circle, show that $\Delta(\epsilon)/\epsilon^h$ is real.

8. If the Alexander polynomial $\Delta(t)$ of a knot is of degree $2h$ and ω is a primitive cube root of unity, show that $\Delta(\omega)/\omega^h$ is an integer.

9. (See Exercise 8.) Show that $\Delta(\omega)/\omega^h = \Delta(\omega^2)/\omega^{2h}$, and hence that $\Delta(\omega)\Delta(\omega^2)$ is the square of an integer.

10. Prove similarly that if i is a primitive fourth root of unity, then $\Delta(i)\Delta(-i)$ is the square of an integer.

APPENDIX I

Differentiable knots are tame. Let K be a knot in 3-dimensional space R^3 which is rectifiable and which is given as the image of a periodic vector-valued function $p(s) = (x(s), y(s), z(s))$ of arc length s whose derivative $p'(s) = (x'(s), y'(s), z'(s))$ exists and is continuous for all s. The period l is the length of K. We shall prove that K is tame, i.e., equivalent to a polygonal knot.

We denote the norm, or length, of a vector $p \in R^3$ by $\| p \|$ and the dot product of two vectors $p_1, p_2 \in R^3$ by $p_1 \cdot p_2$. If neither p_1 nor p_2 is zero, the angle between them is given by

$$\measuredangle (p_1,p_2) = \cos^{-1} \frac{p_1 \cdot p_2}{\| p_1 \| \, \| p_2 \|} .$$

Consider any three parameter values s_0, s_1, s_2 which satisfy $s_0 \leq s_1 < s_2$. From

$$\int_{s_1}^{s_2} p'(u) \, du = \int_{s_1}^{s_2} p'(s_0) \, du + \int_{s_1}^{s_2} (p'(u) - p'(s_0)) \, du$$

follows

$$p(s_2) - p(s_1) = (s_2 - s_1)(p'(s_0) + Q), \tag{1}$$

where

$$Q = \frac{1}{s_2 - s_1} \int_{s_1}^{s_2} (p'(u) - p'(s_0)) \, du.$$

Since parametrization is made with respect to arc length, we have $\| p'(s_0) \| = 1$. Hence,

$$| 1 - \| Q \| | \leq \| p'(s_0) + Q \| \leq 1 + \| Q \|,$$

and so

$$\| p'(s_0) + Q \| = 1 + q$$

for some number q which satisfies $| q | \leq \| Q \|$. Thus,

$$\| p(s_2) - p(s_1) \| = (s_2 - s_1)(1 + q). \tag{2}$$

Choose an arbitrary positive $\epsilon \leq \frac{1}{2}$. Since the derivative $p'(s)$ is continuous and hence uniformly continuous, there exists $\delta > 0$ so that if $| s - s' | < \delta$, then $\| p'(s) - p'(s') \| < \epsilon$. Accordingly, we impose the restriction $s_2 - s_0 < \delta$. Then,

$$(s_2 - s_1) \| Q \| = \| \int_{s_1}^{s_2} (p'(u) - p'(s_0)) \, du \|$$

$$\leq (s_2 - s_1)\epsilon,$$

and

$$| q | \leq \| Q \| \leq \epsilon. \tag{3}$$

Dividing (1) by (2), we obtain

$$\frac{p(s_2) - p(s_1)}{\| p(s_2) - p(s_1) \|} = p'(s_0) + P, \tag{4}$$

where

$$P = \frac{Q - qp'(s_0)}{1 + q}.$$

Since $q \geq -\epsilon \geq -\frac{1}{2}$, we have $1/(1 + q) \leq 2$. Hence,

$$\| P \| \leq 2 \| Q - qp'(s_0) \| \leq 2(\| Q \| + | q |) \leq 4\epsilon. \tag{5}$$

We shall draw two conclusions from the equations of the preceding paragraph. The first, an immediate corollary of (2) and (3), is the well-known fact

(I.1) *The ratio of chord length to arc length along K approaches* 1 *as the latter approaches* 0.

The second conclusion is the principal lemma on which our proof of the tameness of K depends.

(I.2) *For any angle $\alpha > 0$, there exists $\delta > 0$ such that, for any s, s', u, u' in an interval of length δ and such that $s < s'$ and $u < u'$,*

$$\measuredangle\, (p(s') - p(s), p(u') - p(u)) < \alpha.$$

Proof. This lemma is a consequence of (4) and (5). For if

$$s_0 = \min \{s, s', u, u'\},$$

then

$$\frac{p(s') - p(s)}{\| p(s') - p(s) \|} = p'(s_0) + P, \qquad \| P \| \leq 4\epsilon,$$

$$\frac{p(u') - p(u)}{\| p(u') - p(u) \|} = p'(s_0) + P', \qquad \| P' \| \leq 4\epsilon.$$

Hence,

$$\frac{p(s') - p(s)}{\| p(s') - p(s) \|} \cdot \frac{p(u') - p(u)}{\| p(u') - p(u) \|} = 1 + \bar{q},$$

where

$$\bar{q} = p'(s_0) \cdot (P + P') + P \cdot P'.$$

Consequently,

$$| \bar{q} | \leq \| P \| + \| P' \| + \| P \| \| P' \| \leq 8\epsilon + 16\epsilon^2,$$

which can be made arbitrarily small. Thus, $\cos \measuredangle\, (p(s') - p(s), p(u') - p(u))$ can be made arbitrarily near 1, and (I.2) follows.

We now turn to the main argument that K is tame. For any two points $p, p' \in K$, let arc (p,p') be the shorter arc length between them along K. Note that if $| s - s' | \leq l/2$, where l is the total length of the knot, then arc $(p(s),p(s')) = | s - s' |$. Consider the function $f\colon K \times K \to R$ defined by

$$f(p,p') = \begin{cases} \| p - p' \| / \text{arc}\,(p,p'), & p \neq p', \\ 1, & p = p'. \end{cases}$$

We have shown that the ratio of chord length to arc length approaches 1 as the latter approaches 0. Consequently, f is continuous. Since it is positive and its domain is compact, it has a positive minimum value m. Thus,

$$(6) \qquad \| p - p' \| \geq m \,\text{arc}\,(p,p'), \qquad p, p' \in K.$$

We next select a positive angle $\alpha_0 < \pi/4$ such that $\tan \alpha_0 < m/2$. For this angle α_0, choose δ in accordance with Lemma (I.2). Let n be a positive integer so large that $l/n < \delta/2$, and select parameter values $\{s_i\}_{i=-\infty}^{\infty}$ such that $s_{i+1} - s_i = l/n$. Notice that $p(s_i) = p(s_j)$ if and only if $i \equiv j \pmod n$, so that the set $\{p(s_i)\}_{i=-\infty}^{\infty}$ consists of exactly n points of the knot. For each s_i, we form the double solid cone C_i with apex angle α_0 whose axis is the chord joining $p(s_i)$ and $p(s_{i+1})$ (cf. Figure 62).

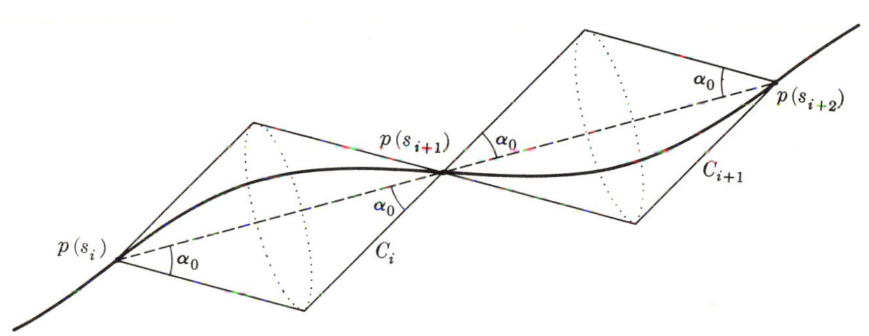

Figure 62

The following four propositions are corollaries of (I.2) and (6).

(I.3) *Adjacent cones intersect only at their common apex.*

Proof. Since $s_{i+2} - s_i = (s_{i+2} - s_{i+1}) + (s_{i+1} - s_i) < \delta$ it follows that the acute angle between the axes of the cones C_i and C_{i+1} is less than α_0, which in turn is less than $\pi/4$. The apex angle of the cones is α_0. Thus, there is no chance of intersection except at the apex.

(I.4) *If $s_i \leq s \leq s_{i+1}$, then $p(s) \in C_i$.*

Proof. We have

$$\measuredangle\,(p(s) - p(s_i),\, p(s_{i+1}) - p(s_i)) < \alpha_0,$$
$$\measuredangle\,(p(s_{i+1}) - p(s),\, p(s_{i+1}) - p(s_i)) < \alpha_0,$$

whence (I.4) follows immediately.

(I.5) *For every perpendicular cross section D of any cone C_i, there is exactly one s in the interval $[s_i, s_{i+1}]$ such that $p(s) \epsilon D$.*

Proof. The existence of s follows from (I.4) and the continuity of the function p. To get uniqueness, suppose that $s_i \leq s < s' \leq s_{i+1}$ and that $p(s)$ and $p(s')$ lie on a single plane perpendicular to the axis of C_i. Then the angle $\sphericalangle (p(s') - p(s), p(s_{i+1}) - p(s_i))$ must, on the one hand, be equal to $\pi/2$ and, on the other hand, be less than α_0. This is a contradiction.

(I.6) *Nonadjacent cones are disjoint.*

Proof. Suppose otherwise, i.e., we assume that there exist nonadjacent cones C_i and C_j and a point p in their intersection, cf. Figure 63. Let $p(s)$ be

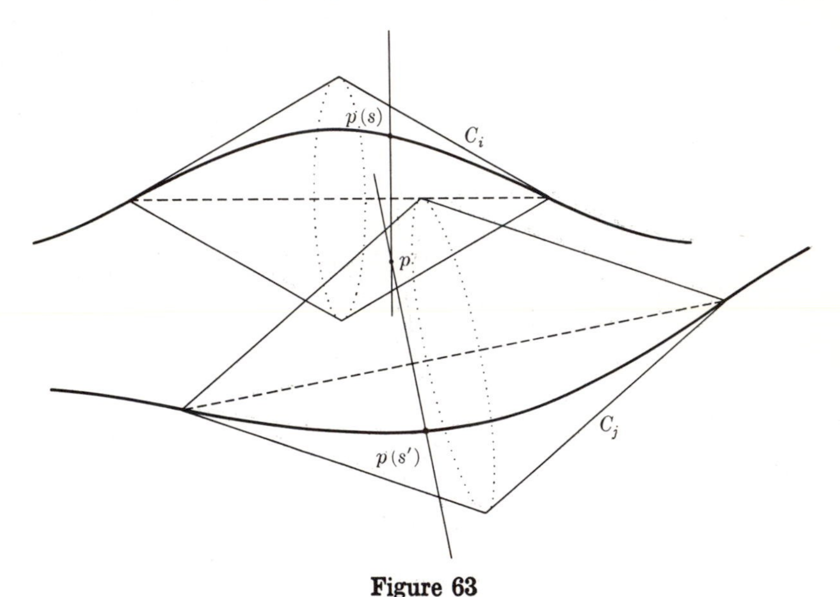

Figure 63

the point with $s_i \leq s \leq s_{i+1}$ on the plane which contains p and is normal to the axis of C_i. The analogous point for C_j is $p(s')$. Then,

$$\| p(s) - p \| \leq \left(\frac{l}{n}\right) \tan \alpha_0,$$

and the same inequality holds for $p(s')$. Since C_i and C_j are not adjacent along K, we know that $\frac{l}{n} \leq \operatorname{arc}(p(s), p(s'))$. Thus,

$$\| p(s) - p(s') \| \leq 2 \operatorname{arc}(p(s), p(s')) \tan \alpha_0 < m \operatorname{arc}(p(s), p(s')).$$

This contradicts (6), and (I.6) is proved.

The proof that K is tame is virtually complete. It only remains to verify that, for each double cone C_i, there exists a homeomorphism h_i of C_i onto itself which is the identity on the boundary of the cone and maps $K \cap C_i$ onto the axis. In view of (I.4) and (I.5), the construction of such a mapping is not hard. Consider an arbitrary closed circular disc D with center p_0. We include the possibility that D is degenerate, i.e., $D = \{p_0\}$. For every interior point p of D, a mapping $g_{D,p}$: $D \rightarrow D$ is defined by mapping any ray joining p to a point q on the circumference of D linearly onto the ray joining p_0 to q so that $p \rightarrow p_0$ and $q \rightarrow q$ (cf. Figure 64). It is obvious that $g_{D,p}$ is a homeo-

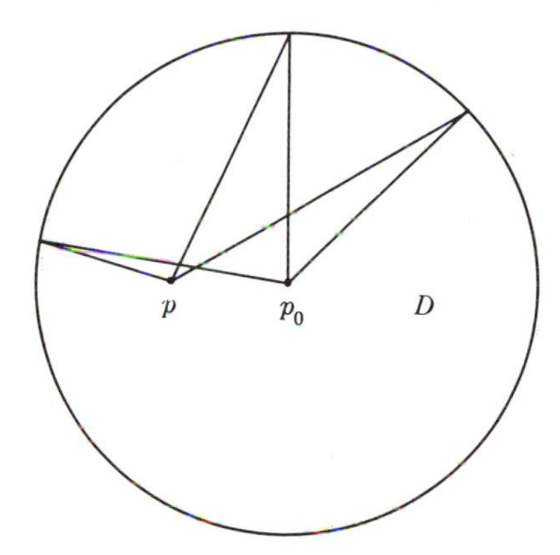

Figure 64. Perpendicular cross section of a cone

morphism of D onto itself which leaves the circumference fixed and maps p onto p_0. Furthermore, $g_{D,p}(p')$ is simultaneously continuous in p and p'.

Returning to the double cone, we consider an arbitrary point $p \in C_i$. Let $p(s)$ be the intersection of the knot K with the plane containing p and normal to the axis of C_i. This plane intersects C_i in a disc (degenerate at the endpoints) which we denote by D_s. The desired homeomorphism h_i: $C_i \rightarrow C_i$ is now defined by

$$h_i(p) = g_{D_s,p(s)}(p).$$

The existence and uniqueness of $p(s)$ as an interior point of D_s are consequences of (I.5) and the proof of (I.4). The final step is the extension of the homeomorphisms h_i to a single mapping h of R_3 onto itself which is defined by

$$h(p) = \begin{cases} p & \text{if} \quad p \notin \bigcup_i C_i, \\ h_i(p) & \text{if} \quad p \in C_i. \end{cases}$$

That h is a well-defined homeomorphism follows from (I.3) and (I.6) and the fact that the homeomorphisms h_i are the identity on the boundaries of the cones. We conclude

(I.7) THEOREM. *The knot K is tame.*

There are two interesting ramifications of this theorem which are worth mentioning. The first is that the cones C_i can clearly be chosen arbitrarily small, i.e., so that the maximum diameter is less than any preassigned ϵ. As a result, the knot K is what is called ϵ-equivalent to a polygonal knot. *For any $\epsilon > 0$, there is a homeomorphism h of R^3 onto itself so that hK is polygonal and $\| h(p) - p \| < \epsilon$ for all $p \in R^3$.* Furthermore, h moves only points lying within a distance ϵ of the knot. The second remark is that *h is realizable by an isotopic deformation of R^3.* This is simply because the mapping $g_{D,p}$ is isotopic to the identity. Using vector notation, we may set

$$g_{D,p,t} = g_{D,t(p-p_0)+p_0}.$$

Thus differentiable knots (as defined in the first paragraph of the appendix) are tame in the strongest possible sense.

The question of when a knot is tame has been studied by several authors. For example Milnor[1] defines the *total curvature* κ of an arbitrarily closed curve and proves (among other results) that if the total curvature of a knot is finite, the knot is tame. He also shows that if a closed curve C is given as the image of a function $p(s)$ of arc length s with continuous 2nd derivative, then κ is given by the usual integral formula

$$\kappa = \int_C \| p''(s) \| \, ds.$$

[1] J. W. Milnor, "On the Total Curvature of Knots," *Ann. of Math.* Vol. 52 (1950), pp. 248–257.

APPENDIX II

Categories and groupoids. The tendency of modern mathematics to isolate almost any set of properties from its original context, to name, and to develop an abstract theory has produced an amazing vocabulary and array of definitions. Obviously, these definitions differ widely in the scope of their applicability and in the depth of the concomitant abstract theories. A few, like that of a group or of a topological space, have a fundamental importance to the whole of mathematics that can hardly be exaggerated. Others are more in the nature of convenient, and often highly specialized, labels which serve principally to pigeonhole ideas. As far as this book is concerned, the notions of category and groupoid belong to the latter class.[1] It is an interesting curiosity that they provide a convenient systematization of the ideas involved in developing the fundamental group.

A set C is a *category* if, for some pairs of elements α, β in C, a product $\alpha \cdot \beta$ in C is defined which satisfies Axioms (*i*) and (*ii*) below. An element ϵ in C is an *identity* if, for any α in C, whenever $\epsilon \cdot \alpha$ (or $\alpha \cdot \epsilon$) is defined, then $\epsilon \cdot \alpha = \alpha$ (or $\alpha \cdot \epsilon = \alpha$).

(*i*) *The product* $\alpha \cdot (\beta \cdot \gamma)$ *is defined if and only if* $(\alpha \cdot \beta) \cdot \gamma$ *is defined. When either is defined the associative law holds*:

$$\alpha \cdot (\beta \cdot \gamma) = (\alpha \cdot \beta) \cdot \gamma.$$

Furthermore, $\alpha \cdot \beta \cdot \gamma$ *is defined (parentheses are dropped by virtue of associativity) if and only if both products* $\alpha \cdot \beta$ *and* $\beta \cdot \gamma$ *are defined.*

(*ii*) *For any element* α *in* C, *there exist identities* ϵ_1 *and* ϵ_2 *in* C *such that* $\epsilon_1 \cdot \alpha$ *and* $\alpha \cdot \epsilon_2$ *are defined.*

The reader will recognize that we have come across these properties before. In fact a fair amount of the material in the beginning of Chapter II can be summarized in the following: *Both the set of all paths in a topological space* X *and the fundamental groupoid* $\Gamma(X)$ *are categories. The mapping of the former category into the latter which assigns to each path* a *its equivalence class* $[a]$ *is onto, product-preserving, and carries the identities of one category onto those of the other.*

We may now observe how the constructions carried out in our development of the fundamental group may be paralleled in abstracto. We prove first

(II.1) *For each* α *in the category* C, *the identities* ϵ_1 *and* ϵ_2 *such that* $\epsilon_1 \cdot \alpha$ *and* $\alpha \cdot \epsilon_2$ *are defined, are unique.*

---

[1] The idea of a category plays a basic role in the axiomatic development of homology theory. In fact the definition and Lemma (II.1) above are taken directly from Chapter IV of S. Eilenberg and N. Steenrod, *Foundations of Algebraic Topology* (Princeton University Press; Princeton, N.J., 1952).

Proof. Suppose there exist identities ϵ_1 and ϵ_1' in C such that $\epsilon_1 \cdot \alpha$ and $\epsilon_1' \cdot \alpha$ are defined. Then $\epsilon_1' \cdot (\epsilon_1 \cdot \alpha)$ is defined since $\epsilon_1 \cdot \alpha = \alpha$. Consequently, $\epsilon_1' \cdot \epsilon_1$ is defined, and hence $\epsilon_1' = \epsilon_1' \cdot \epsilon_1 = \epsilon_1$. Similarly ϵ_2 is unique.

For any category C and identity ϵ in C, we denote by C_ϵ the set of all elements α in C such that $\epsilon \cdot \alpha$ and $\alpha \cdot \epsilon$ are defined.

(II.2) *C_ϵ is a semi-group with identity ϵ.*

Proof. The product of any identity with itself is always defined. Hence, $\epsilon \in C_\epsilon$. Consider next arbitrary elements α and β in C_ϵ. Since $\alpha \cdot \epsilon$ and $\epsilon \cdot \beta$ are defined, $(\alpha \cdot \epsilon) \cdot \beta$ is defined, and $(\alpha \cdot \epsilon) \cdot \beta = \alpha \cdot \beta$. Thus, the product of any two elements of C_ϵ is defined. Since the associative law is known to hold, the proof is complete.

The set of all p-based loops in X and the fundamental group $\pi(X,p)$ are, of course, both examples of sets C_ϵ in their respective categories.

An element α^{-1} in a category C is an *inverse* of an element α if there exist identities ϵ_1 and ϵ_2 in C such that $\alpha \cdot \alpha^{-1} = \epsilon_1$ and $\alpha^{-1} \cdot \alpha = \epsilon_2$.

(II.3) *The products $\epsilon_1 \cdot \alpha$, $\alpha \cdot \epsilon_2$, $\epsilon_2 \cdot \alpha^{-1}$, and $\alpha^{-1} \cdot \epsilon_1$ are defined.*

Proof. Since $\epsilon_1 \cdot \epsilon_1$ is defined, the products $\epsilon_1 \cdot \alpha \cdot \alpha^{-1}$ and $\alpha \cdot \alpha^{-1} \cdot \epsilon_1$ are defined. It follows that $\epsilon_1 \cdot \alpha$ and $\alpha^{-1} \cdot \epsilon_1$ are defined. The analogous argument holds for $\alpha \cdot \epsilon_2$ and $\epsilon_2 \cdot \alpha^{-1}$.

(II.4) *If an inverse exists, it is unique.*

Proof. Suppose β and β' are inverses of α. It follows from (II.3) and (II.1) that the identities ϵ_1 and ϵ_2 whose existence follows from the assumption of the existence of an inverse of α are uniquely determined by α. Consequently, we have

$$\alpha \cdot \beta = \alpha \cdot \beta' = \epsilon_1 \text{ and } \beta \cdot \alpha = \beta' \cdot \alpha = \epsilon_2.$$

Since, by (II.3), $\epsilon_2 \cdot \beta'$ is defined, we have

$$\beta' = \epsilon_2 \cdot \beta' = \beta \cdot \alpha \cdot \beta' = \beta \cdot \epsilon_1 = \beta,$$

and we are done.

A *groupoid* is a category in which every element has an inverse. In view of (2.4) of Chapter II, it is apparent that the set $\Gamma(X)$ of equivalence classes of paths in X does satisfy the requirements of being a groupoid.

(II.5) *If C is a groupoid and ϵ is any identity, then C_ϵ is a group.*

Proof. Consider any α in C_ϵ. Since C is a groupoid, α has an inverse α^{-1}, and there exist identities ϵ_1 and ϵ_2 in C such that

$$\alpha \cdot \alpha^{-1} = \epsilon_1, \qquad \alpha^{-1} \cdot \alpha = \epsilon_2,$$

and the products $\epsilon_1 \cdot \alpha$, $\alpha \cdot \epsilon_2$, $\epsilon_2 \cdot \alpha^{-1}$, and $\alpha^{-1} \cdot \epsilon_1$ are defined. Since $\alpha \in C_\epsilon$, however, the products $\epsilon \cdot \alpha$ and $\alpha \cdot \epsilon$ are defined, and we may therefore conclude from (II.1) that $\epsilon = \epsilon_1 = \epsilon_2$. Consequently, $\alpha^{-1} \in C_\epsilon$. This conclusion, together with (II.2), completes the proof.

The abstract parallel of Theorem (3.1), Chapter II, also holds. The proof is in all essentials identical:

(II.6) *Suppose C is a groupoid, ϵ_1 and ϵ_2 are any two identities in C, and α is an arbitrary element of C such that $\epsilon_1 \cdot \alpha$* and *$\alpha \cdot \epsilon_2$ are defined. Then, for any β in C_{ϵ_1}, the product $\alpha^{-1} \cdot \beta \cdot \alpha$ is defined, and the assignment $\beta \to \alpha^{-1} \cdot \beta \cdot \alpha$ is an isomorphism of C_{ϵ_1} onto C_{ϵ_2}.*

Proof of the van Kampen theorem. This is stated in theorem (3.1) of Chapter V. There are two things to be proved:

(III.1) *The image groups* $\omega_i G_i$, $i = 0, 1, 2$, *generate* G.

Proof. Consider an arbitrary non-trivial element $\alpha \in G$ and a p-based loop $a\colon [0, \| a \|] \to X$ representing α. Since $\alpha \neq 1$, we know that $\| a \| > 0$. We construct a subdivision

$$0 = t_0 < t_1 < \cdots < t_n = \| a \|$$

such that[1] each difference $t_i - t_{i-1}$ is contained in at least one of the inverse images $a^{-1}X_i$, $i = 0, 1, 2$. We then choose an index function μ mapping the integers $1, \cdots, n$ onto 0, 1, 2 such that

$$a[t_{i-1}, t_i] \subset X_{\mu(i)}, \qquad i = 1, \cdots, n.$$

For each point t_i, $i = 0, \cdots, n$, of the subdivision we select a path b_i in X subject to the conditions:

(*i*) $b_i(0) = p$ *and* $b_i(\| b_i \|) = a(t_i)$.

(*ii*) *If* $a(t_i) = p$, *then* $b_i(t) = p$ *for all* t.

(*iii*) $b_i(t) \in X_{\mu(i)} \cap X_{\mu(i+1)}$, $\quad 0 \leq t \leq \| b_i \|$ *and* $i = 1, \cdots, n - 1$.

Notice that Condition (*iii*) may be satisfied because $X_{\mu(i)} \cap X_{\mu(i+1)}$ is one of the subspaces X_0, X_1, and X_2, and each of these is pathwise-connected. Next, consider paths $a_i\colon [0, t_i - t_{i-1}] \to X$, $i = 1, \cdots, n$, defined by $a_i(t) = a(t + t_{i-1})$. Clearly

$$a = \prod_{i=1}^{n} a_i.$$

Since each product $b_{i-1} \cdot a_i \cdot b_i^{-1}$ is defined and b_0 and b_n are identity paths,

$$a \simeq \prod_{i=1}^{n} b_{i-1} \cdot a_i \cdot b_i^{-1}.$$

Each path $b_{i-1} \cdot a_i \cdot b_i^{-1}$ is a p-based loop whose image lies entirely in $X_{\mu(i)}$ and which, therefore, is a representative loop of $\omega_{\mu(i)}\alpha_i$ for some $\alpha_i \in G_{\mu(i)}$. Thus

$$\alpha = \prod_{i=1}^{n} \omega_{\mu(i)}\alpha_i,$$

and the proof of (III.1) is complete.

[1] S. Lefschetz, *Algebraic Topology* (American Mathematical Society Colloquium Publications Vol. 27; New York, 1942), p. 37.

The second part of the van Kampen theorem is

(III.2) *If H is an arbitrary group and $\psi_i\colon G_i \to H$, $i = 0, 1, 2$, are homomorphisms which satisfy $\psi_0 = \psi_1\theta_1 = \psi_2\theta_2$, then there exists a unique homomorphism $\lambda\colon G \to H$ such that $\psi_i = \lambda\omega_i$, $i = 0, 1, 2$.*

Proof. The uniqueness of λ is no problem. If it exists, the relations $\psi_i = \lambda\omega_i$, $i = 0, 1, 2$, together with the conclusion of (III.1) imply that it is unique. The only question is of existence, and there is an obvious construction. Let α be an arbitrary element of G. We have shown that

$$\alpha = \prod_{i=1}^{n} \omega_{\mu(i)}\alpha_i.$$

So we define

$$\lambda\alpha = \prod_{i=1}^{n} \psi_{\mu(i)}\alpha_i.$$

The hard problem is to prove that λ is well-defined. If it is, we are finished; for the preceding formula implies both that λ is a homomorphism and that it satisfies $\psi_i = \lambda\omega_i$, $i = 0, 1, 2$. The problem clearly amounts to proving that, for any finite set of elements $\alpha_i \in G_{\mu(i)}$, where $i = 1, \cdots, r$ and μ is any mapping of the integers $1, \cdots, r$ into $0, 1, 2$, then

$$\prod_{i=1}^{r} \omega_{\mu(i)}\alpha_i = 1 \text{ implies } \prod_{i=1}^{r} \psi_{\mu(i)}\alpha_i = 1.$$

Verification of this proposition is the objective of the remainder of the proof.

We select representative loops $a_i \in \alpha_i$, $i = 1, \cdots, r$. Then the product

$$a = \prod_{i=1}^{r} \omega_{\mu(i)}a_i$$

is equivalent to the identity path. (For simplicity we shall denote an inclusion mapping and its induced homomorphism of the fundamental groups by the same symbol.) The equivalence is effected by a fixed-endpoint family $\{h_s\}$ or, what amounts to the same thing, a continuous mapping $h\colon R \to X$, where

$$R = [0, \| a \|] \times [0, 1],$$

which satisfies

$$h(t,0) = a(t),$$
$$h(0,s) = h(t,1) = h(\| a \|, s) = p.$$

The vertical lines $t = \sum_{k=1}^{i} \| a_k \|$, $i = 1, \cdots, r$, provide a decomposition of the rectangle R, and we consider a refinement

$$0 = t_0 < t_1 < \cdots < t_n = \| a \|,$$
$$0 = s_0 < s_1 < \cdots < s_m = 1,$$

into subrectangles R_{ij}, so fine[1] that each R_{ij} is contained in at least one of the inverse images $h^{-1}X_i$, $i = 0, 1, 2$. Each subrectangle R_{ij} consists of all pairs (t,s) satisfying the inequalities $t_{i-1} \leq t \leq t_i$ and $s_{j-1} \leq s \leq s_j$, $i = 1, \cdots, n$ and $j = 1, \cdots, m$. The subdivision has been chosen so fine that there exists an index function $\nu(i,j)$ such that

$$hR_{ij} \subset X_{\nu(i,j)}, \qquad i = 1, \cdots, n \quad \text{and} \quad j = 1, \cdots, m.$$

For each lattice point (t_i,s_j), we select a path e_{ij} in X subject to the conditions:

(*iv*) *The initial and terminal points of* e_{ij} *are* p *and* $h(t_i,s_j)$, *respectively.*

(*v*) *If* $h(t_i,s_j) = p$, *then* $e_{ij}(t) = p$ *for all* t.

(*vi*) *The image of the path* e_{ij} *is contained in*

$$X_{\nu(i,j)} \cap X_{\nu(i+1,j)} \cap X_{\nu(i,j+1)} \cap X_{\nu(i+1,j+1)}.$$

(Assume $X_{\nu(i,j)} = X$ if $i = 0$ or $n + 1$, or if $j = 0$ or $m + 1$.)

(*vii*) *If* $\sum_{k=1}^{j-1} \| a_k \| \leq t_{i-1} < t_i \leq \sum_{k=1}^{j} \| a_k \|$, *then the image of* e_{i0} *is a subset of* $X_{\mu(j)}$.

Conditions (*iv*), (*v*), and (*vi*) are entirely analogous to (*i*), (*ii*), and (*iii*),

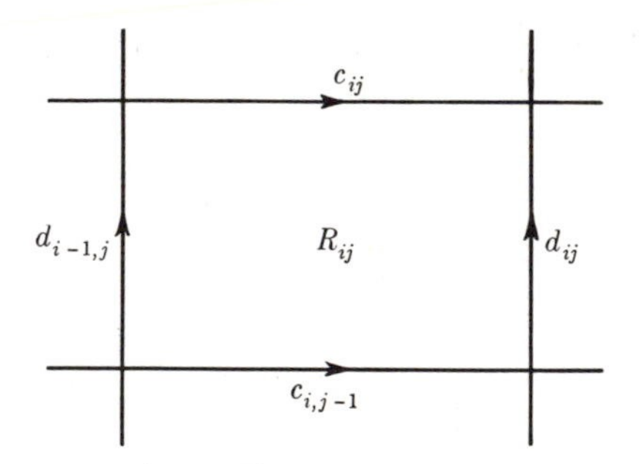

Figure 65

respectively; (*vii*) is an additional complexity. We next define paths, (cf. Figure 65)

$$c_{ij}(t) = h(t + t_{i-1}, s_j), \qquad 0 \leq t \leq t_i - t_{i-1},$$
$$d_{ij}(s) = h(t_i, s + s_{j-1}), \qquad 0 \leq s \leq s_j - s_{j-1},$$

and set

$$a_{ij} = e_{i-1,j} \cdot c_{ij} \cdot e_{ij}^{-1}, \qquad i = 1, \cdots, n \quad \text{and} \quad j = 0, \cdots, m,$$
$$b_{ij} = e_{i,j-1} \cdot d_{ij} \cdot e_{ij}^{-1}, \qquad i = 0, \cdots, n \quad \text{and} \quad j = 1, \cdots, m.$$

It is a consequence of (*vi*) that the image points of the loops a_{ij}, b_{ij}, $a_{i,j-1}$, and $b_{i-1,j}$ all lie in $X_{\nu(i,j)}$. Hence, they define group elements α_{ij}, β_{ij}, α_{ij}', and β_{ij}', respectively, in $G_{\nu(i,j)}$. The product $a_{i,j-1} \cdot b_{ij} \cdot a_{ij}^{-1} \cdot b_{i-1,j}^{-1}$ is contractible (i.e., equivalent to the identity path) in X; moreover, since the image of R_{ij} as well as the images of the four paths lies in $X_{\nu(i,j)}$, the product is also contractible in $X_{\nu(i,j)}$. We conclude that

$$\alpha_{ij}'\beta_{ij}\alpha_{ij}^{-1}(\beta_{ij}')^{-1} = 1. \tag{1}$$

The central idea in the proof of (III.2) is the fact that *if group elements* $\alpha \in G_i$ *and* $\beta \in G_j$, $i,j = 0, 1, 2$, *possess a common representative loop, then* $\psi_i\alpha = \psi_j\beta$.

The proof is easy: Since $X_i \cap X_j = X_k$ for some $k = 0$, 1, or 2, each of the two inclusion mappings

$$X_i \xleftarrow{\eta_1} X_k \xrightarrow{\eta_2} X_j$$

is either an identity mapping or one of θ_1 and θ_2. As a result, the induced homomorphisms

$$G_i \xleftarrow{\eta_1} G_k \xrightarrow{\eta_2} G_j$$

must be consistent with the homomorphisms ψ_0, ψ_1, and ψ_2, i.e.,

$$\psi_i\eta_1 = \psi_k = \psi_j\eta_2.$$

The assertion that α and β possess a common representative loop states that there exists a p-based loop c in X_k such that $\eta_1 c \in \alpha$ and $\eta_2 c \in \beta$. Thus, if c defines $\gamma \in G_k$, we have

$$\eta_1\gamma = \alpha, \qquad \eta_2\gamma = \beta.$$

Hence,

$$\psi_i\alpha = \psi_i\eta_1\gamma = \psi_k\gamma = \psi_j\eta_2\gamma = \psi_j\beta,$$

and the assertion is proved.

Applying this result, we obtain

$$\begin{cases} \psi_{\nu(i,j)}\alpha_{ij} = \psi_{\nu(i,j+1)}\alpha'_{i,j+1}, \\ \psi_{\nu(i,j)}\beta_{ij} = \psi_{\nu(i+1,j)}\beta'_{i+1,j}. \end{cases} \tag{2}$$

Now apply the homomorphism $\psi_{\nu(i,j)}$ to (1). The equation obtained says that the result of reading counterclockwise around each R_{ij} under $\psi_{\nu(i,j)}$ is the identity. Equations (2) show that edges of adjacent rectangles will cancel. It follows (by induction) that the result of reading around the circumference of the large rectangle R is the identity. Furthermore, only the elements along the bottom edge, $s = 0$, are nontrivial. We conclude, therefore, that

$$\prod_{i=1}^{n} \psi_{\nu(i,1)}\alpha_{i1}' = 1.$$

Since each of the numbers $\sum_{k=1}^{j} \| a_k \|, j = 1, \cdots, r$, is a member of $\{t_1, \cdots, t_n\}$, there exists an index function $i(j)$ such that $i(0) = 0$ and

$$t_{i(j)} = \sum_{k=1}^{j} \| a_k \|, \qquad j = 1, \cdots, r.$$

Then, as a result of Conditions (*iv*) and (*v*), we have

$$\prod_{i=i(j-1)+1}^{i(j)} a_{i0} \simeq \omega_{\mu(j)} a_j, \qquad j = 1, \cdots, r.$$

By virtue of (*vii*), we may assume that the equivalence is in $X_{\mu(j)}$. Thus, each loop a_{i0}, $i = i(j-1)+1, \cdots, i(j)$, determines a group element $\alpha_i' \in G_{\mu(j)}$ and

$$\prod_{i=i(j-1)+1}^{i(j)} \alpha_i' = \alpha_j.$$

Since α_{i1}' and α_i' possess a common representative loop a_{i0}, it follows from our central assertion that

$$\psi_{\mu(j)}\alpha_i' = \psi_{\nu(i,1)}\alpha_{i1}', \qquad i = i(j-1)+1, \cdots, i(j).$$

Finally, therefore,

$$1 = \prod_{j=1}^{r}\left(\prod_{i=i(j-1)+1}^{i(j)} \psi_{\nu(i,1)}\alpha_{i1}'\right) = \prod_{j=1}^{r}\left(\prod_{i=i(j-1)+1}^{i(j)} \psi_{\mu(j)}\alpha_i'\right)$$
$$= \prod_{j=1}^{r} \psi_{\mu(j)}\left(\prod_{i=i(j-1)+1}^{i(j)} \alpha_i'\right) = \prod_{j=1}^{r} \psi_{\mu(j)}\alpha_j,$$

and the proof of (III.2) is complete.

The above proof of the van Kampen theorem can be used to prove a more general theorem,[2] from which the present proof is virtually copied. Instead of regarding X as the union of just two subspaces X_1 and X_2, we consider an arbitrary collection of pathwise-connected, open subsets X_i (i may range over any index set whatever) which is closed under finite intersections and which satisfies

$$X = \bigcup X_i$$
$$p \in \bigcap X_i, \quad \text{for some } p.$$

Let $G_i = \pi(X_i,p)$ and $G = \pi(X,p)$ and consider all homomorphisms θ_{ij}: $G_i \to G_j$ and ω_i: $G_i \to G$ induced by inclusion (the existence of θ_{ij} presupposes that $X_i \subset X_j$). Then, the conclusions of the van Kampen theorem hold: *The groups $\omega_i G_i$ generate G, and, for any group H and homomorphisms ψ_i: $G_i \to H$ which satisfy $\psi_i = \psi_j\theta_{ij}$, there exists a unique homomorphism λ: $G \to H$ such that $\psi_i = \lambda\omega_i$.*

This generalization may be used to calculate the fundamental group of the union of an increasing nest of open sets each of whose groups is known. This result can be used to obtain presentations of the groups of wild knots and other wild imbeddings.[3]

[2] R. H. Crowell, "On the van Kampen theorem," *Pacific J. Math.* Vol. 9, No. 1 (1959), pp. 43–50.

[3] R. H. Fox and E. Artin, "Some Wild Cells and Spheres in Three-dimensional Space." *Ann. of Math.* Vol. 49 (1948), pp. 979–990.

Guide to the Literature

The literature of knot theory is scattered, and some of it is difficult reading. The only comprehensive book on the subject is [Reidemeister 1932], and the literature has more than tripled since then. The following notes are intended to help the student find some of the more easily accessible papers and to orient him in the field. For the most part, the papers quoted are recent ones. The references are to the subsequent Bibliography, which is a chronological listing. Such references as [Fox 1954], [Brody 1960′], and [Murasugi 1958″] refer respectively to the first, second, and third paper of the author within the year indicated. Many important earlier papers that are not quoted in the Guide can be found in the Bibliography and in the bibliographies of the quoted papers.

The problem with which we have been concerned in this book is a special case of the *problem of placement:* Given topological spaces X and Y, what are the different ways of imbedding X in Y? The case that we have studied is $X = S^1$, $Y = R^3$. Its significance is that it is the simplest interesting case and that the methods used to study it have, *mutatis mutandis*, general validity.

Thus we may always consider the *group* $G = \pi(Y - X)$ of a placement $X \subset Y$. If this group is finitely presented it has *Alexander matrices* and *elementary ideals* [Fox 1954], but if the group cannot be finitely presented its Alexander matrices are infinite matrices and things get more complicated [Brody 1960′].

If X is a *link* of μ components, i.e., the union of μ mutually disjoint simple closed curves in $Y = R^3$, the commutator quotient group G/G' is free abelian of rank μ, so that we must deal with L-polynomials in μ variables. If the link is tame, a polynomial $\Delta(t_1, \cdots, t_\mu)$ can be defined even when $\mu \geq 2$, and it has properties analogous to the Alexander polynomial in one variable [Reidemeister and Schumann 1934, Fox 1954, Torres and Fox 1954, Hosokawa 1958, Fox 1960′]. If X is just any one-dimensional complex in $Y = R^3$, then the group G/G', though free abelian, may no longer have a preferred basis, and this causes special difficulties [Kinoshita 1958′, 1959].

A natural generalization of knot theory is the case X an m-sphere, or union of μ (≥ 2) mutually disjoint m-spheres, and $Y = R^n$ ($n > m$). It is reasonably well-established that the case of a single m-sphere (knotting) is really interesting only if $m = n - 2$, while the case of several m-spheres (linking) is not interesting unless $\frac{n-1}{2} \leq m \leq n-2$ [M. Brown 1960, Stallings 1961, Zeeman 1960]. If $m = n - 2$, G/G' is free abelian of rank μ, and the theory is very similar to that of knots and links in R^3 [Artin 1925′, van Kampen 1928, Andrews and Curtis 1959, Fox and Milnor 1957, Terasaka 1959, Zeeman 1960, Kinoshita 1961], however for $n > 3$ the reciprocal character of the polynomial Δ no longer holds in general. Naturally the case $n = 4$ has received the most attention. If X is a surface other than S^2 in R^4, G/G' may have elements of finite order, and this causes new difficulties [Fox 1960′, Kinoshita 1961′]. Knots and links in arbitrary 3-dimensional manifolds have been considered [Blanchfield 1957, Brody 1960] although much remains to be done. This is an especially interesting case because of the possibility of applying it to the yet unsolved problem of classifying the 3-manifolds. The group itself is a more powerful invariant than the polynomial, so naturally less is known about its properties [Fox 1948, Torres and Fox 1954, Rapport 1960, Neuwirth 1959].

Placement of surfaces, with or without boundary, in R^3 has some interesting and difficult problems that are relevant to knot theory [Alexander 1924, Fox 1948, Kyle 1955]. Especial attention should be paid to the *Dehn lemma*, whose solution was one of topology's recent important breakthroughs [Papakyriakopoulos 1957, Shapiro and Whitehead 1958, Papakyriakopoulos 1958]. An immediate consequence of Dehn's lemma is the fact that a tame knot (in R^3) is trivial if and only if its group is cyclic [Dehn 1910].

A simple and elegant construction shows that a tame knot can always be spanned by an orientable surface, and this fact can be used to give an especially practical form of the Alexander matrix [Seifert 1934, Wendt 1937, Fox 1960].

One of the most important chapters of knot theory has to do with covering spaces. Unbranched covering spaces are described, for example, in the book of Seifert and Threlfall 1934, and the description of branched ones has been recently formalized [Fox 1957]. Every closed orientable 3-manifold is a branched covering space of S^3 [Alexander 1919 together with Clifford 1877]. By means of the branched cyclic covering spaces of a knot (or link) new geometric meaning can be attached to the various aspects of the Alexander matrix, and even more powerful invariants can be defined [Seifert 1933′, 1935, Blanchfield and Fox 1951, Kyle 1954, 1959, Fox 1956, 1960].

The theory of *companionship* of knots includes the *multiplication* (composition), *doubling*, and *cabling* of knots [Schubert 1953, 1954, Whitehead 1937, Seifert 1949]. To multiply two knots you simply tie one after the other in the same piece of string. Under this operation the tame knot types form a commutative semigroup S in which factorization is unique [Schubert 1949]. In this semigroup only the trivial type has an inverse; this proves that it is impossible to tie two knots in a piece of string in such a way that they 'cancel'. The problem "Which knot types can appear when a (locally flat) S^2 in R^4

is cut by a hyperplane R^3?" leads to a classification of knot types such that these classes of types, with multiplication induced from S, form a group [Fox and Milnor 1957, Terasaka 1959]. Not much is known, as yet, about this group.

There are interesting problems of knot theory centered about *knot-diagrams*, i.e., regular projections of knots. Thus *alternating knots*, i.e., those that have projections whose crossings are alternately under and over around the knot, have some surprising properties [Murasugi 1958, 1958′, 1958″, 1960, Crowell 1959, 1959′], one of which virtually amounts to a pun: the Alexander polynomial of an alternating knot is alternating. The problem of recognizing from their diagrams whether two knots are equivalent is in general unsolved, but a method has recently been given for deciding from a diagram whether a knot is trivial [Haken 1961], and the solutions of similar problems for alternating knots and links were already known [Crowell 1959]. If a knot has a diagram in which there is only one overpass it is obviously trivial; those knots that have diagrams containing just two overpasses have been completely classified [Schubert 1956]. It is known that any tame knot has a certain kind of diagram called a *plat* [Reidemeister 1960]; the simplest case is that of a plat with four strings (Viergeflechte) and this has received some attention [Bankwitz and Schumann 1934].

The homotopy groups π_k ($k = 1, 2, \cdots$) generalize the fundamental group $\pi = \pi_1$, so it is natural to examine the homotopy groups of $Y - X$ for a placement of X in Y. It is now known [Papakyriakopoulos 1957] that $\pi_k(S^3 - X)$ is trivial for $k \geq 2$ for any tame knot X, and that if X is a link $\pi_2(S^3 - X)$ is trivial if and only if X cannot be "pulled apart into two pieces". An earlier investigation into this problem led to some highly interesting algebraic problems but no general solution [Whitehead 1939, Higman 1948]. If X is an $(n - 2)$-sphere in S^n or a union of mutually disjoint $(n - 2)$-spheres in S^n, then $\pi_2(S^n - X)$ may or may not be trivial and some interesting problems arise [Andrews and Curtis 1959, Epstein 1960].

The most venerable invariant of knot theory is the linking number of a link of two components; this was first considered over a hundred years ago [Gauss 1833]. Its value can be read from a diagram [Brunn 1892] or from its polynomial [Reidemeister and Schumann 1934, Torres and Fox 1954]. It has been generalized in various ways that deserve further study [Pannwitz 1933, Eilenberg 1937, Milnor 1954, 1957, Plans 1957].

Can the set of fixed points of a transformation of R^3 of finite period p be a (tame) knot? This problem is unsolved, although some results on it have been obtained [Montgomery and Samelson 1955, Kinoshita 1958′, Fox 1958]. A related problem concerns the knots that can be mapped on themselves by transformations of period p [Trotter 1961], and it is also only slightly solved.

The connections between knot theory and differential geometry [Fáry 1949, Milnor 1950, 1953, Fox 1950] and between knot theory and algebraic geometry [Zariski 1935, Reeve 1955] deserve further exploration.

There is a bewildering array of papers on wild knots and on wildness, of which we can indicate only a sampling [Borsuk 1947, Fox and Artin 1948, Fox 1949, Blankinship and Fox 1950, Kirkor 1958, 1958′, Bing 1958, Debrunner and Fox 1960, Brody 1960′].

Finally we must mention the closely related theory of braids [Artin 1925, 1947, 1947′, 1950, Newman 1942, Markoff 1935, Weinberg 1939, Fox and Neuwirth 1962] and several provocative papers of mysterious significance [Fox 1958′, Kinoshita and Terasaka 1957, Hashizume and Hosokawa 1958, Curtis 1959].

The standard table of (prime) knots of 9 or fewer crossings may be found in the book of Reidemeister, 1932, pp. 25, 31, 41, 70–72. The tables on pp. 70–72 were extended by various workers in the 19th century up through 10 crossings and through the alternating 11 crossings. The corresponding extension of the table on p. 41 of $\Delta(t)$ has been made by machine but has not yet been published. No corresponding tables of links have ever been made.

The Ashley book of knots [Ashley 1944] is an immense compendium of knots, as the term is understood by sailors, weavers, etc. With a little patience one can find in it all sorts of provocative examples of knots and links.

Bibliography

This bibliography contains only the most important papers on knot theory. References to papers that are false, trivial, or of specialized or peripheral interest have been suppressed. Abbreviations J.F.M., Zbl., and M.R. are for the titles of the reviewing journals *Jahrbuch über die Fortschritte der Mathematik*, *Zentralblatt für Mathematik*, and *Mathematical Reviews*.

1833 Gauss, K. F. "Zur mathematischen Theorie der electrodynamischen Wirkungen." *Werke. Königlichen Gesellschaft der Wissenschaften zu Göttingen*. 1877, vol. 5, p. 605.

1877 Clifford, W. K. "On the canonical form and dissection of a Riemann's surface." *Proceedings of the London Mathematical Society*, vol. 8, pp. 292–304; J.F.M. 9, 391.

1877–1885 Tait, P. G. "On knots." *Scientific Paper, I.* (Cambridge University Press, 1898, London), pp. 273–347; J.F.M. 9, 392; 11, 362; 17, 521.

1890 Little, C. N. "Alternate $\pm$ knots of order 11." *Trans. Roy. Soc. Edinburgh*, vol. 36, pp. 253–255 (with two plates); J.F.M. 22, 562.

1892 Brunn, H. "Topologische Betrachtungen." *Zeitschrift für Mathematik und Physik*, vol. 37, pp. 106–116; J.F.M. 24, 507.

1900 Little, C. N. "Non-alternate $\pm$ knots." *Trans. Roy. Soc. Edinburgh*, vol. 39, pp. 771–778 (with three plates); J.F.M. 31, 481.

1910 Dehn, M. "Über die Topologie des drei-dimensionalen Raumes." *Mathematische Annalen*, vol. 69, pp. 137–168; J.F.M. 41, 543.

1914 Dehn, M. "Die beiden Kleeblattschlingen." *Math. Ann.*, vol. 75, pp. 402–413.

1919 Alexander, J. W. "Note on Riemann spaces." *Bulletin of the American Mathematical Society*, vol. 26, pp. 370–372. J.F.M. 47, 529.

1921 Antoine, L. "Sur l'homéomorphie de deux figures et de leurs voisinages." *Journal de Mathématique*, series 8, vol. 4, pp. 221–325; J.F.M. 48, 650.

1923 Schreier, O. "Über die Gruppen $A^aB^b = 1$." *Abhandlungen aus dem Mathematischen Seminar der Universität, Hamburg*, vol. 3, pp. 167–169; J.F.M. 50, 70.

1924 Alexander, J. W. "On the subdivision of 3-space by a polyhedron." *Proceedings of the National Academy of Sciences of the United States of America*, vol. 10, pp. 6–8; J.F.M. 50, 659.

Alexander, J. W. "An example of a simply connected surface bounding a region which is not simply connected." *Proc. Nat. Acad.*, vol. 10, pp. 8–10; J.F.M. 50, 661.

Alexander, J. W. "Remarks on a point set constructed by Antoine." *Proc. Nat. Acad.*, vol. 10, pp. 10–12; J.F.M. 50, 661.

1925 Artin, E. "Theorie der Zöpfe." *Hamburg Abh.*, vol. 4, pp. 47–72; J.F.M. 51, 450.

Artin, E. "Zur Isotopie zweidimensionaler Flächen im R_4." *Hamburg Abh.* vol. 4, pp. 174–177; J.F.M. 51, 450.

1926 Reidemeister, K. "Knoten und Gruppen." *Hamburg Abh.*, vol. 5, pp. 7–23; J.F.M. 52, 578.

Reidemeister, K. "Elementare Begründung der Knotentheorie." *Hamburg Abh.*, vol. 5, pp. 24–32; J.F.M. 52, 579.

Reidemeister, K. "Über unendliche diskrete Gruppen." *Hamburg Abh.*, vol. 5, pp. 33–39; J.F.M. 52, 112.

Schreier, O. "Die Untergruppen der freien Gruppen." *Hamburg Abh.*, vol. 5, pp. 233–244; J.F.M. 53, 110.

1927 Alexander, J. W. and Briggs, G. B. "On types of knotted curves." *Annals of Mathematics*, series 2, vol. 28, pp. 562–586; J.F.M. 53, 549.

1928 Alexander, J. W. "Topological invariants of knots and links." *Transactions of the American Mathematical Society*, vol. 30, pp. 275–306; J.F.M. 54, 603.

Reidemeister, K. "Über Knotengruppen." *Hamburg Abh.*, vol. 6, pp. 56–64; J.F.M. 54, 603.

van Kampen, E. R. "Zur Isotopie zweidimensionaler Flächen im R_4." *Hamburg Abh.*, vol. 6, p. 216; J.F.M. 54, 602.

1929 Kneser, H. "Geschlossene Flächen in dreidimensionalen Mannigfaltigkeiten." *Jahresbericht der Deutschen Mathematiker Vereinigung*, vol. 38, pp. 248–260; J.F.M. 55^1, 311.

Reidemeister, K. "Knoten und Verkettungen." *Mathematische Zeitschrift*, vol. 29, pp. 713–729; J.F.M. 55^2, 973.

1930 Bankwitz, C. "Über die Torsionszahlen der zyklischen Überlagerungsräume des Knotenraumes." *Ann. of Math.*, vol. 31, pp. 131–133; J.F.M. 56^2, 1133.

Frankl, F. and Pontrjagin, L. "Ein Knotensatz mit Anwendung auf die Dimensionstheorie." *Math. Ann.*, vol. 102, pp. 785–789; J.F.M. 56^1, 503.

1931 Magnus, W. "Untersuchungen über einige unendliche diskontinuierliche Gruppen." *Math. Ann.*, vol. 105, pp. 52–74; J.F.M. 57^1, 151.

1932 Alexander, J. W. "Some problems in topology." *Verhandlungen des Internationallen Mathematiker-Kongress Zürich*, vol. 1, pp. 249-257; J.F.M. 58^1, 621; Zbl. 6, 421.

Burau, W. "Über Zopfinvarianten." *Hamburg Abh.*, vol. 9, pp. 117–124; J.F.M. 58^1, 614; Zbl. 6, 34.

Burau, W. "Kennzeichnung der Schlauchknoten." *Hamburg Abh.*, vol. 9, pp. 125–133; J.F.M. 58^1, 615; Zbl. 6, 34.

Reidemeister, K. KNOTENTHEORIE in *Ergebnisse der Mathematik*, vol. 1, no. 1 (reprint Chelsea, 1948, New York); J.F.M. 58^2, 1202; Zbl. 5, 120.

1933 Goeritz, L. "Knoten und quadratische Formen." *Math. Z.*, vol. 36, pp. 647–654; J.F.M. 59^2, 1237; Zbl. 6, 422.

Pannwitz, E. "Eine elementargeometrische Eigenschaft von Verschlingungen und Knoten." *Math. Ann.*, vol. 108, pp. 629–672; J.F.M. 59^1, 557; Zbl. 7, 231.

Seifert, H. "Topologie dreidimensionaler gefaserter Räume." *Acta Mathematica*, vol. 60, pp. 147–238; J.F.M. 59^2, 1241; Zbl. 6, 83.

Seifert, H. "Verschlingungsinvarianten." *Sitzungsberichte der preussischen Akademie der Wissenschaften*, vol. 26, pp. 811–828; J.F.M. 59^2, 1238; Zbl. 8, 181.

van Kampen, E. R. "On the connection between the fundamental groups of some related spaces." *American Journal of Mathematics*, vol. 55, pp. 261–267; J.F.M. 59^1, 577; Zbl. 6, 415.

1934 Bankwitz, C. and Schumann, H. G. "Über Viergeflechte." *Hamburg Abh.*, vol. 10, pp. 263–284; J.F.M. 60^1, 527; Zbl. 9, 230.

Burau, W. "Kennzeichnung der Schlauchverkettungen." *Hamburg Abh.*, vol. 10, pp. 285–297; J.F.M. 60^1, 525; Zbl. 9, 231.

Goeritz, L. "Die Betti'schen Zahlen der zyklischen Überlagerungsräume der Knotenaussenräume." *Amer. J. Math.*, vol. 56, pp. 194–198; J.F.M. 60^1, 524; Zbl. 9, 39.

Goeritz, L. "Bemerkungen zur Knotentheorie." *Hamburg Abh.*, vol. 10, pp. 201–210; J.F.M. 60^1, 525; Zbl. 9, 230.

Reidemeister, K. and Schumann, H. G. "L-Polynome von Verkettungen." *Hamburg Abh.*, vol. 10, pp. 256–262; J.F.M. 60^1, 526; Zbl. 9, 230.

Seifert, H. "Über das Geschlecht von Knoten." *Math. Ann.*, vol. 110, pp. 571–592; J.F.M. 60^1, 523; Zbl. 10, 133.

Seifert, H. and Threlfall, W. LEHRBUCH DER TOPOLOGIE. (Teubner, 1934, Leipzig und Berlin); J.F.M. 60^1, 496; Zbl. 9, 86.

1935 Bankwitz, C. "Über Knoten und Zöpfe in gleichsinniger Verdrillung." *Math. Z.*, vol. 40, pp. 588–591; J.F.M. 61^2, 1352.

Burau, W. "Über Verkettungsgruppen." *Hamburg Abh.*, vol. 11, pp. 171–178; J.F.M. 61^2, 1021; Zbl. 11, 177.

Burau, W. "Über Zopfgruppen und gleichsinnig verdrillte Verkettungen." *Hamburg Abh.*, vol. 11, pp. 179–186; J.F.M. 61^1, 610; Zbl. 11, 178.

Markoff, A. "Über die freie Äquivalenz geschlossener Zöpfe." *Recueil Mathématique Moscou*, Vol. 1, pp. 73–78; J.F.M. 62^1, 658; Zbl. 14, 42.

Seifert, H. "Die Verschlingungsinvarianten der zyklischen Knotenüberlagerungen." *Hamburg Abh.*, vol. 11, pp. 84–101; J.F.M. 61^1, 609; Zbl. 11, 178.

Zariski, O. ALGEBRAIC SURFACES in *Ergebnisse der Mathematik*, vol. 3, no. 5, Springer, Berlin (reprinted Chelsea, 1948, New York); J.F.M. 61^1, 704.

1936 Eilenberg, S. "Sur les courbes sans noeuds." *Fundamenta Mathematicae* vol. 28, pp. 233–242; Zbl. 16, 138.

Seifert, H. "La théorie des noeuds." *L'enseignement Math.*, vol. 35, pp. 201–212; J.F.M. 62^1, 659; Zbl. 15, 84.

1937 Eilenberg, S. "Sur les espaces multicohérents, II." *Fund. Math.*, vol. 29, pp. 101–122; Zbl. 17, 40.

Newman, M. H. A. and Whitehead, J. H. C. "On the group of a certain linkage." *Quarterly Journal of Mathematics* (Oxford series), vol. 8, pp. 14–21; J.F.M. 63^1, 552; Zbl. 16, 278.

Wendt, H. "Die gordische Auflösung von Knoten." *Math. Z.*, vol. 42, pp. 680–696; J.F.M. 63^1, 552; Zbl. 16, 420.

Whitehead, J. H. C. "On doubled knots." *Journal of the London Mathematical Society*, vol. 12, pp. 63–71; J.F.M. 63^1, 552; Zbl. 16, 44.

1939 Vietoris, L. "Über m-gliedrige Verschlingungen." *Jahresbericht Deutsche Mathematiker Vereinigung*, vol. 49, Abt. 1, pp. 1–9; Zbl. 20, 407.

Weinberg, N. "Sur l'equivalence libre des tresses fermées." *Comptes Rendus* (*Doklady*) *de l'Academie des Sciences de l'URSS*, vol. 23, pp. 215–216; Zbl. 21, 357.

Whitehead, J. H. C. "On the asphericity of regions in a 3-sphere." *Fund. Math.* vol. 32, pp. 149–166; Zbl. 21, 162.

1942 Newman, M. H. A. "On a string problem of Dirac." *J. London Math. Soc.*, vol. 17, pp. 173–177; Zbl. 28, 94; M.R. 4, 252.

Tietze, H. EIN KAPITEL TOPOLOGIE. *Zur Einführung in die Lehre von den verknoteten Linien*. Teubner, Leipzig und Berlin (*Hamburger Mathematische Einzelschriften* 36); M. R. 8, 285.

1944 Ashley, C. W. THE ASHLEY BOOK OF KNOTS. Doubleday and Co., N.Y.

1947 Artin, E. "Theory of braids." *Ann. of Math.*, vol. 48, pp. 101–126; Zbl. 30, 177; M.R. 8, 367.

Artin, E. "Braids and permutations." *Ann. of Math.*, vol. 48, pp. 643–649; Zbl. 30, 178; M.R. 9, 6.

Bohnenblust, F. "The algebraical braid group." *Ann. of Math.*, vol. 48, pp. 127–136; Zbl. 30, 178; M.R. 8, 367.

Borsuk, K. "An example of a simple arc in space whose projection in every plane has interior points." *Fund. Math.*, vol. 34, pp. 272–277; Zbl. 32, 314; M.R. 10, 54.

1948 Borsuk, K. "Sur la courbure totale des courbes fermées." *Annales de la société Polonaise de mathématique*, vol. 20, pp. 251–265; M.R. 10, 60.

Chow, W. L. "On the algebraical braid group." *Ann. of Math.*, vol. 49, pp. 654–658; Zbl. 33, 10; M.R. 10, 98.

Fox, R. H. "On the imbedding of polyhedra in 3-space." *Ann. of Math.*, vol. 49, pp. 462–470; Zbl. 32, 125; M.R. 10, 138.

Fox, R. H. and Artin, E. "Some wild cells and spheres in three-dimensional space." *Ann. of Math.* vol. 49, pp. 979–990; Zbl. 33, 136; M.R. 10, 317.

Higman, G. "A theorem on linkages." *Quart. J. of Math.* (*Oxford series*), vol. 19, pp. 117–122; Zbl. 30, 322; M.R. 9, 606.

1949 Burger, E. "Über Schnittzahlen von Homotopie-ketten." *Math. Z.*, vol. 52, pp. 217–255; Zbl. 33, 307; M.R. 12, 43.

Fary, I. "Sur la courbure totale d'une courbe gauche faisant un noeud." *Bulletin de la Société Mathématique de France*, vol. 77, pp. 128–138; M.R. 11, 393.

Fox, R. H. "A remarkable simple closed curve." *Ann. of Math.*, vol. 50, pp. 264–265; Zbl. 33, 136; M.R. 11, 45.

Schubert, H. "Die eindeutige Zerlegbarkeit eines Knotens in Primknoten." *Sitzungsberichte der Heidelberger Akademie der Wissenschaften Mathematisch-Naturwissenschaftliche Klasse*, No. 3, pp. 57–104; Zbl. 31, 286; M.R. 11, 196.

Seifert, H. "Schlingknoten." *Math. Z.*, vol. 52, pp. 62–80; Zbl. 33, 137; M.R. 11, 196.

1950 Artin, E. "The theory of braids." *American Scientist*, vol. 38, pp. 112–119; M.R. 11, 377.

Blankinship, W. A. and Fox, R. H. "Remarks on certain pathological open subsets of 3-space and their fundamental groups." *Proceedings of the American Mathematical Society*, vol. 1, pp. 618–624; Zbl. 40, 259; M.R. 13, 57.

Burger, E. "Über Gruppen mit Verschlingungen." *Journal für die Reine und Angewandte Mathematik*, vol. 188, pp. 193–200; Zbl. 40, 102; M.R. 13, 204.

Fox, R. H. "On the total curvature of some tame knots." *Ann. of Math.* vol. 52, pp. 258–260; Zbl. 37, 390; M.R. 12, 373.

Graeub, W. "Die semilinearen Abbildungen." *S.-B. Heidelberger Akad. Wiss. Math. Nat. kl.*, pp. 205–272; M.R. 13, 152.

Milnor, J. W. "On the total curvature of knots." *Ann. of Math.*, vol. 52, pp. 248–257; Zbl. 37, 389; M.R. 12, 373.

Seifert, H. "On the homology invariants of knots." *Quart. J. Math. Oxford*, vol. 1, pp. 23–32; Zbl. 35, 111; M.R. 11, 735.

Seifert, H. and Threlfall, W. "Old and new results on knots." *Canadian Journal of Mathematics*, vol. 2, pp. 1–15; Zbl. 35, 251; M.R. 11, 450.

1951 Blanchfield, R. C. and Fox, R. H. "Invariants of self-linking." *Ann. of Math.*, vol. 53, pp. 556–564; Zbl. 45, 443; M.R. 12, 730.

Blankinship, W. A. "Generalization of a construction of Antoine." *Ann. of Math.*, vol. 53, pp. 276–297; Zbl. 42, 176; M.R. 12, 730.

Chen, K. T. "Integration in free groups." *Ann. of Math.*, vol. 54, pp. 147–162; Zbl. 45, 301; M.R. 13, 105.

Torres, G. "Sobre las superficies orientables extensibles en nudos." *Boletin de la Sociedad Matemática Mexicana*, vol. 8, pp. 1–14; M.R. 13, 375.

1952 Chen, K. T. "Commutator calculus and link invariants." *Proc. A.M.S.* vol. 3, pp. 44–55; Zbl. 49, 404; M.R. 13, 721.

Chen, K. T. "Isotopy invariants of links." *Ann. of Math.*, vol. 56, pp. 343–353; Zbl. 49, 404; M.R. 14, 193.

Fox, R. H. "On the complementary domains of a certain pair of inequivalent knots." *Koninklijke Nederlandse Akademie van Wetenschappen. Proceedings*, series A, vol. 55 (or equivalently, *Indagationes Mathematicae ex Actis Quibus Titulis* vol. 14), pp. 37–40; Zbl. 46, 168; M.R. 13, 966.

Fox, R. H. "Recent development of knot theory at Princeton." *Proceedings of the International Congress of Mathematics*, Cambridge, 1950, vol. 2, pp. 453–457; Zbl. 49, 130; M.R. 13, 966.

Moise, E. E. "Affine structures in 3-manifolds, V. The triangulation theorem and Hauptvermutung." *Ann. of Math.*, vol. 56, pp. 96–114; Zbl. 48, 171; M.R. 14, 72.

1953 Fox, R. H. "Free differential calculus, I. Derivation in the free group ring." *Ann. of Math.*, vol. 57, pp. 547–560; Zbl. 50, 256.

Gugenheim, V. K. A. M. "Piecewise linear isotopy and embedding of elements and spheres. I, II." *Proc. Lond. Math. Soc.*, series 2, vol. 3, pp. 29–53, 129–152; Zbl. 50, 179; M.R. 15, 336.

Kneser, M. and Puppe, D. "Quadratische Formen und Verschlingungsinvarianten von Knoten." *Math. Z.*, vol. 58, pp. 376–384; Zbl. 50, 398; M.R. 15, 100.

Milnor, J. "On the total curvatures of closed space curves." *Mathematica Scandinavica*, vol. 1, pp. 289–296; Zbl. 52, 384; M.R. 15, 465.

Plans, A. "Aportación al estudio de los grupos de homología de los recubrimientos ciclicos ramificados correspondientes a un nudo." *Revista de la Real Academía de Ciencias Exactas, Fisicas y Naturales de Madrid*, vol. 47, pp. 161–193; Zbl. 51, 146; M.R. 15, 147.

Schubert, H. "Knoten und Vollringe." *Acta Math.*, vol. 90, pp. 131–286; Zbl. 51, 404; M.R. 17, 291.

Torres, G. "On the Alexander polynomial." *Ann. of Math.*, vol. 57, pp. 57–89; Zbl. 50, 179; M.R. 14, 575.

1954 Bing, R. H. "Locally tame sets are tame." *Ann. of Math.*, vol. 59, pp. 145–158; Zbl. 55, 168; M.R. 15, 816.

Fox, R. H. "Free differential calculus, II. The isomorphism problem." *Ann. of Math.*, vol. 59, pp. 196–210; M.R. 15, 931.

Homma, T. "On the existence of unknotted polygons on 2-manifolds in E^3." *Osaka Mathematical Journal*, vol. 6, pp. 129–134; Zbl. 55, 421; M.R. 16, 160.

Kyle, R. H. "Branched covering spaces and the quadratic forms of a link." *Ann. of Math.*, vol. 59, pp. 539–548; Zbl. 55, 421; M.R. 15, 979.

Milnor, J. "Link groups." *Ann. of Math.*, vol. 59, pp. 177–195; Zbl. 55, 169; M.R. 17, 70.

Moise, E. E. "Affine structures in 3-manifolds VII, invariance of the knot types; local tame imbedding." *Ann. of Math.*, vol. 59, pp. 159–170; Zbl. 55, 168; M.R. 15, 889.

Schubert, H. "Über eine numerische Knoteninvariante." *Math. Z.*, vol. 61, pp. 245–288; Zbl. 58, 174; M.R. 17, 292.

Torres, G. and Fox, R. H. "Dual presentations of the group of a knot." *Ann. of Math.*, vol. 59, pp. 211–218; Zbl. 55, 168; M.R. 15, 979.

1955 Kyle, R. H. "Embeddings of Möbius bands in 3-dimensional space." *Proceedings of the Royal Irish Academy*, Section A, vol. 57, pp. 131–136; Zbl. 66, 171; M.R. 19, 976.

Montgomery, D. and Samelson, H. "A theorem on fixed points of involutions in S^3." *Canadian J. Math.*, vol. 7, pp. 208–220; Zbl. 64, 177; M.R. 16, 946.

Papakyriakopoulos, C. D. "On the ends of knot groups." *Ann. of Math.*, vol. 62, pp. 293–299; Zbl. 67, 158; M.R. 19, 976.

Reeve, J. E. "A summary of results in the topological classification of plane algebroid singularities." *Rendiconti del Seminario Matematico di Torino*, vol. 14, pp. 159–187.

1956 Aumann, R. J. "Asphericity of alternating knots." *Ann. of Math.*, vol. 64, pp. 374–392; Zbl. 78, 164; M.R. 20, 453.

Bing, R. H. "A simple closed curve that pierces no disk." *Journal de Mathématiques Pures et Appliquées series 9*, vol. 35, pp. 337–343; Zbl. 70, 402; M.R. 18, 407.

Fox, R. H. "Free differential calculus, III. Subgroups." *Ann. of Math.*, vol. 64, pp. 407–419; M.R. 20, 392.

Schubert, H. "Knoten mit zwei Brücken." *Math. Z.*, vol. 65, pp. 133–170; Zbl. 71, 390; M.R. 18, 498.

1957 Bing, R. H. "Approximating surfaces with polyhedral ones." *Ann. of Math.*, vol. 65, pp. 456–483; M.R. 19, 300.

Blanchfield, R. C. "Intersection theory of manifolds with operators with applications to knot theory." *Ann. of Math.*, vol. 65, pp. 340–356; Zbl. 80, 166; M.R. 19, 53.

Conner, P. E. "On the action of a finite group on $S^n \times S^n$." *Annals of Math.*, vol. 66, pp. 586–588; M.R. 20, 453.

Fox, R. H. "Covering spaces with singularities." *Lefschetz symposium. Princeton Mathematical Series*, vol. 12, pp. 243–257; Zbl. 79, 165; M.R. 23, 106.

Fox, R. H. and Milnor, J. W. "Singularities of 2-spheres in 4-space and equivalence of knots." *Bull. A. M. S.*, vol. 63, p. 406.

Kinoshita, S. and Terasaka, H. "On unions of knots." *Osaka Math. J.*, vol. 9, pp. 131–153; Zbl. 80, 170; M.R. 20, 804.

Milnor, J. "Isotopy of links." *Lefschetz symposium. Princeton Math. Ser.*, vol. 12, pp. 280–306; Zbl. 80, 169; M.R. 19, 1070.

Papakyriakopoulos, C. D. "On Dehn's lemma and the asphericity of knots." *Proc. Nat. Acad. Sci. U.S.A.*, vol. 43, pp. 169–172; *Ann. of Math.*, vol. 66, pp. 1–26; Zbl. 78, 164; M.R. 18, 590; 19, 761.

Plans, A. "Aportación a la homotopía de sistemas de nudos." *Revista Matemática Hispano-Americana,* Series 4, vol. 17, pp. 224–237; M.R. 20, 803.

1958 Bing, R. H. "Necessary and sufficient conditions that a 3-manifold be S^3." *Ann. of Math.*, vol. 68, pp. 17–37; Zbl. 81, 392; M.R. 20, 325.

Fox, R. H. "On knots whose points are fixed under a periodic transformation of the 3-sphere." *Osaka Math. J.*, vol. 10, pp. 31–35; Zbl. 84, 395.

Fox, R. H. "Congruence classes of knots." *Osaka Math. J.*, vol. 10, pp. 37–41; Zbl. 84, 192.

Fox, R. H., Chen, K. T., and Lyndon, R. C. "Free differential calculus, IV. The quotient groups of the lower central series." *Ann. of Math.*, vol. 68, pp. 81–95; M.R. 21, 247.

Hashizume, Y. "On the uniqueness of the decomposition of a link." *Osaka Math. J.*, vol. 10, pp. 283–300, vol. 11, p. 249; M.R. 21, 308.

Hashizume, Y. and Hosokawa, F. "On symmetric skew unions of knots." *Proceedings of the Japan Academy*, vol. 34, pp. 87–91; M.R. 20, 804.

Hosokawa, F. "On ∇-polynomials of links." *Osaka Math. J.*, vol. 10, pp. 273–282; M.R. 21, 308.

Kinoshita, S. "On Wendt's theorem of knots II." *Osaka Math. J.*, vol. 10, pp. 259–261.

Kinoshita, S. "On knots and periodic transformations." *Osaka Math. J.*, vol. 10, pp. 43–52; M.R. 21, 434.

Kinoshita, S. "Alexander polynomials as isotopy invariants, I." *Osaka Math. J.*, vol. 10, pp. 263–271; M.R. 21, 308.

Kirkor, A. "A remark about Cartesian division by a segment." *Bulletin de l'Académie Polonaise des Sciences*, vol. 6, pp. 379–381; M.R. 20, 580.

Kirkor, A. "Wild O-dimensional sets and the fundamental group." *Fund. Math.*, vol. 45, pp. 228–236; Zbl. 80, 168; M.R. 21, 300.

Murasugi, K. "On the genus of the alternating knot, I." *Journal of the Mathematical Society of Japan*, vol. 10, pp. 94–105; M.R. 20, 1010.

Murasugi, K. "On the genus of the alternating knot, II." *J. Math. Soc. Japan*, vol. 10, pp. 235–248; M.R. 20, 1010.

Murasugi, K. "On the Alexander polynomial of the alternating knot." *Osaka Math. J.*, vol. 10, pp. 181–189; M.R. 20, 1010.

Papakyriakopoulos, C. D. "Some problems on 3-dimensional manifolds." *Bull. A. M. S.*, vol. 64, pp. 317–335; M.R. 21, 307.

Rabin, M. O. "Recursive unsolvability of group theoretic problems." *Annals of Math.*, vol. 67, pp. 172–194.

Shapiro, A. and Whitehead, J. H. C. "A proof and extension of Dehn's lemma." *Bull. A. M. S.*, vol. 64, pp. 174–178; M.R. 21, 432.

Whitehead, J. H. C. "On 2-spheres in 3-manifolds." *Bull. A. M. S.*, vol. 64, pp. 161–166; M.R. 21, 432.

1959 Andrews, J. J. and Curtis, M. L. "Knotted 2-spheres in the 4-sphere." *Ann. of Math.*, vol. 70, pp. 565–571; M.R. 21, 1111.

Anger, A. L. "Machine calculation of knot polynomials." *Princeton senior thesis.*

Conner, P. E. "Transformation groups on a K(π, 1)." *Michigan Mathematics Journal*, vol. 6, pp. 413–417; M.R. 23, 113.

Crowell, R. H. "Non-alternating links." *Illinois Journal of Mathematics*, vol. 3, pp. 101–120; M.R. 20, 1010.

Crowell, R. H. "On the van Kampen theorem." *Pacific Journal of Mathematics* vol. 9, pp. 43–50; M.R. 21, 713.

Crowell, R. H. "Genus of alternating link types." *Ann. of Math.*, vol. 69, pp. 258–275; M.R. 20, 1010.

Curtis, M. L. "Self-linked subgroups of semigroups." *Amer. J. Math.*, vol. 81, pp. 889–892; M.R. 21,1343.

Goblirsch, R. P. "On decomposition of 3-space by linkages." *Proc. A. M. S.* 10, pp. 728–730; M.R. 22, 507.

Homma, T. "On tame imbeddings of O-dimensional compact sets in E^3." *Yokohama Mathematics Journal*, vol. 7, pp. 191–195; M.R. 23, 241.

Kinoshita, S. "Alexander polynomials as isotopy invariants, II." *Osaka Math. J.*, vol. 11, pp. 91–94; M.R. 22, 170.

Kyle, R. H. "Branched covering spaces and the quadratic forms of links, II." *Ann. of Math.* vol. 69, pp. 686–699; M.R. 21, 1111.

Neuwirth, L. "Knot groups." *Princeton Ph. D. Thesis.*

Terasaka, H. "On null-equivalent knots." *Osaka Math. J.*, vol. 11, pp. 95–113; M.R. 22, 1447.

1960 Brody, E. J. "The topological classification of the lens spaces." *Ann. of Math.*, vol. 71, pp. 163–184; M.R. 22, 1215.

Brody, E. J. "On infinitely generated modules." *Quart. J. Oxford*, vol. 11, pp. 141–150; M.R. 22, 1701.

Brown, M. "A proof of the generalized Schoenflies theorem." *Bull. A. M. S.*, vol. 66, pp. 74–76; M.R. 22, 1441.

Debrunner, H. and Fox, R. H. "A mildly wild imbedding of an n-frame." *Duke Mathematical Journal*, vol. 27, pp. 425–429; M.R. 22, 1939.

Epstein, D. B. A. "Linking spheres." *Proceedings of the Cambridge Philosophical Society*, vol. 56, pp. 215–219; M.R. 22, 1448.

Fisher, G. M. "On the group of all homeomorphisms of a manifold." *Trans. A. M. S.*, vol. 97, pp. 193–212; M.R. 22, 1443.

Fox, R. H. "The homology characters of the cyclic coverings of the knots of genus one." *Ann. of Math.*, vol. 71, pp. 187–196; M.R. 22, 1702.

Fox, R. H. "Free differential calculus, V. The Alexander matrices reexamined." *Ann. of Math.*, vol. 71, pp. 408–422; M.R. 22, 444.

Hosokawa, F. and Kinoshita, S. "On the homology group of the branched cyclic covering spaces of links." *Osaka Math. J.*, vol. 12, pp. 331–355.

Hotz, G. "Arkadenfadendarstellung von Knoten und eine neue Darstellung der Knotengruppe. *Hamburg Abh.*, vol. 24, pp. 132–148; M.R. 22, 337.

Kinoshita, S. "On diffeomorphic approximations of polyhedral surfaces in 4-space." *Osaka Math. J.*, vol. 12, pp. 191–194.

Murasugi, K. "On alternating knots." *Osaka Math. J.*, vol. 12, pp. 277–303.

Neuwirth, L. "The algebraic determination of the genus of knots." *Amer. J. Math.*, vol. 82, pp. 791–798; M.R. 22, 1946.

Noguchi, H. "The smoothing of combinatorial n-manifolds in $(n + 1)$-space." *Ann. of Math.*, vol. 72, pp. 201–215.

Rapaport, E. S. "On the commutator subgroup of a knot group." *Ann. of Math.*, vol. 71, pp. 157–162; M.R. 22, 1159.

Reidemeister, K. "Knoten und Geflechte." *Akademie der Wissenschaft in Gottingen Mathematisch-physikalische Klasse. Nachrichten . . . Mathematisch-physikalisch-chemische Abteilung*, Vol. 5, pp. 105–115; M.R. 22, 337.

Terasaka, H. "On the non-triviality of some kinds of knots." *Osaka Math. J.*, vol. 12, pp. 113–144.

Terasaka, H. "Musubime no riron." (Japanese) *Sugaku*, vol. 12, pp. 1–20.

Zeeman, E. C. "Unknotting spheres." *Ann. of Math.*, vol. 72, pp. 350–361; M.R. 22, 1447.

Zeeman, E. C. "Linking spheres." *Hamburg Abh.*, vol. 24, pp. 149–153; M.R. 22, 1448.

Zeeman, E. C. "Unknotting spheres in five dimensions." *Bull. A. M. S.*, vol. 66, p. 198; M.R. 22, 1447.

1961 Bing, R. H. "Tame Cantor sets in E^3." *Pacific J. Math.*, vol. 2, pp. 435–446.

Crowell, R. H. "Knots and wheels." *N.C.T.M. Yearbook.*

Crowell, R. H. "Corresponding group and module sequences." *Nagoya Mathematical Journal*, vol. 19, pp. 27–40.

Debrunner, H. "Links of Brunnian type." *Duke Math. J.*, vol. 28, pp. 17–23.

Epstein, D. B. A. "Projective planes in 3-manifolds." *Proc. London Math. Soc.*, vol. 11, pp. 469–484.

Fadell, E. and van Buskirk, J. "On the braid groups of E^2 and S^2." *Bull. A. M. S.*, vol. 67, pp. 211–213.

Gassner, M. J. "On braid groups." *Hamburg Abh.*, vol. 25, pp. 10–12.

Gay, D. A. "A problem involving certain knots of ten and eleven crossings." *Princeton senior thesis.*

Gluck, H. "The embedding of two-spheres in the four-sphere." *Bull. A. M. S.*, vol. 67, pp. 586–589.

Gluck, H. "Orientable surfaces in four-space." *Bull. A. M. S.*, vol. 67, pp. 590–592.

Haken, W. "Theorie der Normalflächen." *Acta Math.*, vol. 105, pp. 245–375.

Keldysh, L. V. "Some problems of topology in Euclidean spaces." *Russian Mathematical Surveys*, vol. 16, pp. 1–15.

Kervaire, J. and Milnor, J. "On 2-spheres in 4-manifolds." *Proc. Nat. Acad.*, vol. 47, pp. 1651–1657.

Kinoshita, S. "On the Alexander polynomial of 2-spheres in a 4-sphere." *Ann. of Math.*, vol. 74, pp. 518–531.

Lipschutz, S. "On a finite matrix representation of the braid group." *Archiv der Mathematik*, vol. 12, pp. 7–12.

Murasugi, K. "On the definition of the knot matrix." *Proc. Japan Acad.*, vol. 37, pp. 220–221.

Murasugi, K. "Remarks on torus knots." *Proc. Japan Acad.*, vol. 37, p. 222.

Murasugi, K. "Remarks on knots with two bridges." *Proc. Japan Acad.*, vol. 37, pp. 294–297.

Neuwirth, L. "The algebraic determination of the topological type of the complement of a knot." *Proc. A. M. S.*, vol. 12, pp. 904–906.

Neuwirth, L. "A note on torus knots and links determined by their groups." *Duke Math. J.*, vol. 28, pp. 545–551.

Schubert, H. "Bestimmung der Primfaktorzerlegung von Verkettungen." *Math. Zeitschrift*, vol. 76, pp. 116–148.

Terasaka, H. and Hosokawa, F. "On the unknotted sphere S^2 in E^4" *Osaka Math. J.*, vol. 13, pp. 265–270.

Trotter, H. "Periodic automorphisms of groups and knots." *Duke Math. J.*, vol. 28, pp. 553–557.

1962 Alford, W. R. "Some "nice" wild 2-spheres in E^3." TOPOLOGY OF

3-MANIFOLDS. *Proceedings of the* 1961 *Topology Institute at the University of Georgia*, Prentice-Hall, pp. 29–33.

Ball, B. J. "Penetration indices and applications." *Proc. Top. Inst.*, pp. 37–39.

Bing, R. H. "Decompositions of E^3." *Proc. Top. Inst.*, pp. 5–21.

Brown, M. "Locally flat imbeddings of topological manifolds." *Ann. of Math.*, vol. 75, pp. 331–341.

Dahm, D. M. "A generalization of braid theory." *Princeton Ph.D. thesis.*

Doyle, P. H. "Tame, wild, and planar sets in E^3." *Proc. Top. Inst.*, pp. 34–36.

Edwards, C. H. "Concentric tori and tame curves in S^3." *Proc. Top. Inst.*, pp. 39–41; M.R. 23, 107.

Fadell, E. and Neuwirth, L. "Configuration spaces". *Math. Scand.*, vol. 10, pp. 111–118.

Fox, R. H. "A quick trip through knot theory." *Proc. Top. Inst.*, pp. 120–167.

Fox, R. H. "Some problems in knot theory." *Proc. Top. Inst.*, pp. 168–176.

Fox, R. H. "Knots and periodic transformations." *Proc. Top. Inst.*, pp. 177–182.

Fox, R. H. and Harrold, O. G. "The Wilder arcs." *Proc. Top. Inst.*, pp. 184–187.

Fox, R. H. "Construction of simply connected 3-manifolds." *Proc. Top. Inst.*, pp. 213–216.

Fox, R. H. and Neuwirth, L. "The braid groups." *Math. Scand.*, vol. 10, pp. 119–126.

Gluck, H. "Rotational symmetry of 3-manifolds." *Proc. Top. Inst.*, pp. 104–106.

Gluck, H. "The reducibility of embedding problems." *Proc. Top. Inst.*, pp. 182–183.

Haefliger, A. "Knotted $(4k - 1)$-spheres in 6k-space." *Ann. of Math.*, vol. 75, pp. 452–466.

Haefliger, A. "Differentiable links." *Topology*, vol. 1, pp. 241–244.

Haken, W. "Ueber das Homöomorphieproblem der 3-Mannigfaltigkeiten I." *Math. Zeitschr.*, vol. 80, pp. 89–120.

Harrold, O. G. "Combinatorial structures, local unknottedness, and local peripheral unknottedness." *Proc. Top. Inst.*, pp. 71–83.

Hempel, J. "Construction of orientable 3-manifolds." *Proc. Top. Inst.*, pp. 207–212.

Kinoshita, S. "On quasi-translations in 3-space." *Proc. Top. Inst.*, pp. 223–226.

Kinoshita, S. "A note on the genus of a knot." *Proc. A.M.S.*, vol. 13, p. 451.

Mazur, B. "Symmetric homology spheres." *Ill. J. Math.*, vol. 6, pp. 245–250.

Milnor, J. "A unique decomposition theorem for 3-manifolds." *Am. J. Math.*, vol. 84, pp. 1–7.

Milnor, J. "A duality theorem for Reidemeister torsion." *Ann. of Math.*, vol. 76, pp. 137–147.

Moise, E. "Periodic homeomorphisms of the 3-sphere." *Ill. J. Math.*, vol. 6, pp. 206–225.

Murasugi, K. "Non-amphicheirality of the special alternating links." *Proc. A.M.S.*, vol. 13, pp. 771–776.

Shepperd, J. A. H. "Braids which can be plaited with their threads tied together at an end." *Proceedings of the Royal Society*, vol. A 265, pp. 229–244.

Stallings, J. "On fibering certain 3-manifolds." *Proc. Top. Inst.*, pp. 95–100.

Stoel, T. B. "An attempt to distinguish certain knots of ten and eleven crossings." *Princeton senior thesis.*

Trotter, H. F. "Homology of group systems with applications to knot theory." *Ann. of Math.*, vol. 76, pp. 464–498.

Yajima, T. "On the fundamental groups of knotted 2-manifolds in the 4-space." *J. Osaka Math.*, vol. 13, pp. 63–71.

Zeeman, E. C. "Isotopies and knots in manifolds." *Proc. Top. Inst.*, pp. 187–193.

Zeeman, E. C. "Unknotting 3-spheres in six dimensions." *Proc. A.M.S.*, vol. 13, pp. 753–757.

1963 Burde, G. "Zur Theorie der Zopfe." *Math. Ann.*, vol. 151, pp. 101–107.

Crowell, R. H. "The group G'/G'' of a knot group G." *Duke Math. J.*, vol. 30, pp. 349–354.

Crowell, R. H. and Trotter, H. F. "A class of pretzel knots." *Duke Math. J.*, vol. 30, pp. 373–377.

Gluck, H. "Unknotting S^1 in S^4." *Bull. A.M.S.*, vol. 69, pp. 91–94.

Haefliger, A. "Plongement differentiable dans le domaine stable." *Comm. Math. Helv.*, vol. 37, pp. 155–176.

Hammer, G. "Ein Verfahren zur Bestimmung von Begleitknoten." *Math. Z.*, vol. 81, pp. 395–413.

Hudson, J. F. P. "Knotted tori." *Topology*, vol. 2, pp. 11–22.

Lipschutz, S. "Note on a paper by Shepperd on the braid group." *Proc. A.M.S.*, vol. 14, pp. 225–227.

Murasugi, K. "On a certain subgroup of the group of an alternating link." *Amer. J. Math.*, vol. 85, pp. 544–550.

Neuwirth, L. "On Stallings fibrations." *Proc. A.M.S.*, vol. 14, pp. 380–381.

Neuwirth, L. "A remark on knot groups with a center." *Proc. A.M.S.*, vol. 14, pp. 378–379.

Neuwirth, L. "Interpolating manifolds for knots in S^3." *Topology*, vol. 2, pp. 359–365.

Noguchi, H. "A classification of orientable surfaces in 4-space." *Proc. Japan Acad.*, vol. 39, pp. 422–423.

Schmid, J. "Ueber eine Klasse von Verkettungen." *Math. Z.*, vol. 81, pp. 187–205.

Stallings, J. "On topologically unknotted spheres." *Ann. of Math.*, vol. 77, pp. 490–503.

Takase, R. "Note on orientable surfaces in 4-space." *Proc. Japan Acad.*, vol. 39, p. 424.

Zeeman, E. "Unknotting combinatorial falls." *Math. Ann.*, vol. 78, pp. 501–520.

1964 Andrews, J. and Dristy, F. "The Minkowski units of ribbon knots." *Proc. A.M.S.*, vol. 15, pp. 856–864.

Bing, R. and Klee, V. "Every simple closed curve in E^3 is unknotted in E^4." *J. London Math. Soc.*, vol. 39, pp. 86–94.

Boardman, J. "Some embeddings of 2-spheres in 4-manifolds." *Proc. Camb. Phil. Soc.*, vol. 60, pp. 354–356.

Crowell, R. H. "On the annihilator of a knot module." *Proc. A.M.S.*, vol. 15, pp. 696–700.

Fox, R. H. and Smythe, N. "An ideal class invariant of knots." *Proc. A.M.S.*, vol. 15, pp. 707–709.

Krotenheerdt, O. "Ueber einen speziellen Typ alternierender Knoten." *Math. Ann.*, vol. 153, pp. 270–284.

Milnor, J. "Most knots are wild." *Fund. Math.*, vol. 54, pp. 335–338.

Murasugi, K. "The center of a group with one defining relation." *Math. Ann.*, vol. 155, pp. 246–251.

Trotter, H. F. "Noninvertible knots exist." *Topology*, vol. 2, pp. 275–280.

Yajima, T. "On simply knotted spheres in R^4." *J. Osaka Math.*, vol. 1, pp. 133–152.

Yanagawa, T. "Brunnian systems of 2-spheres in 4-space." *J. Osaka Math.*, vol. 1, pp. 127–132.

1965 Brown, E. M. and Crowell, R. H. "Deformation retractions of 3-manifolds into their boundaries." *Ann. of Math.*, vol. 82, pp. 445–458.

Crowell, R. H. "Torsion in link modules." *J. Math. Mech.*, vol. 14, pp. 289–298.

Haefliger, A. and Steer, B. "Symmetry of linking coefficients." *Comment. Math. Helv.*, vol. 39, pp. 259–270.

Levine, J. "A characterization of knot polynomials." *Topology*, vol. 4, pp. 135–141.

Levine, J. "A classification of differentiable knots." *Ann. of Math.*, vol. 82, pp. 15–51.

Murasugi, K. "On a certain numerical invariant of link types." *Trans. A.M.S.*, vol. 117, pp. 387–422.

Murasugi, K. "On the Minkowski unit of slice links." *Trans. A.M.S.*, vol. 114, pp. 377–383.

Murasugi, K. "Remarks on rosette knots." *Math. Ann.*, vol. 158, pp. 290–292.

Murasugi, K. "On the center of the group of a link." *Proc. A.M.S.*, vol. 16, pp. 1052–1057.

Neuwirth, L. KNOT GROUPS, *Annals of Mathematics Studies*, No. 56, Princeton University Press, Princeton, N.J.

Robertello, R. "An invariant of knot cobordism." *Comm. Pure Appl. Math.*, vol. 18, pp. 543–555.

Zeeman, E. "Twisting spin knots." *Trans. A.M.S.*, vol. 115, pp. 471–495.

1966 Brown, E. M. and Crowell, R. H. "The augmentation subgroup of a link." *J. Math. Mech.*, vol. 15, pp. 1065–1074.

Burde, G. and Zieschang, H. "Eine Kennzeichnung der Torusknoten." *Math. Ann.*, vol. 167, pp. 169–176.

Feustel, C. D. "Homotopic arcs are isotopic." *Proc. A.M.S.*, vol. 17, pp. 891–896.

Fox, R. H. "Rolling." *Bull. A.M.S.*, vol. 72, pp. 162–164.

Fox, R. H. and Milnor, J. "Singularities of 2-spheres in 4-space and cobordism of knots." *Osaka J. Math.*, vol. 3, pp. 257–267.

Giffen, C. "The generalized Smith conjecture." *Amer. J. Math.*, vol. 88, pp. 187–198.

Holmes, R. and Smythe, N. "Algebraic invariants of isotopy of links." *Amer. J. Math.*, vol. 88, pp. 646–654.

Kinoshita, S. "On Fox's property of a surface in a 3-manifold." *Duke Math. J.*, vol. 33, pp. 791–794.

Levine, J. "Polynomial invariants of knots of codimension two." *Ann. of Math.*, vol. 84, pp. 537–554.

Murasugi, K. "On Milnor's invariants for links." *Trans. A.M.S.*, vol. 124, pp. 94–110.

Schaufele, C. B. "A note on link groups." *Bull. A.M.S.*, vol. 72, pp. 107–110.

Smythe, N. "Boundary links." TOPOLOGY SEMINAR WISCONSIN, *Annals of Mathematics Studies,* No. 60, Princeton University Press, pp. 69–72.

1967 Burde, G. and Zieschang, H. "Neuwirthsche Knoten und Flächenabbildungen," *Abh. Math. Sem. Univ. Hamburg*, vol. 31, pp. 239–246.

Burde, G. "Darstellungen von Knotengruppen." *Math. Ann.*, vol. 173, pp. 24–33.

Fox, R. H. "Two theorems about periodic transformations of the 3-sphere." *Michigan Math. J.*, vol. 14, pp. 331–334.

Gamst, J. "Linearisierung von Gruppendaten mit Anwendungen auf Knotengruppen." *Math. Z.*, vol. 97, pp. 291–302.

Giffen, C. "On transformations of the 3-sphere fixing a knot." *Bull. A.M.S.*, vol. 73, pp. 913–914.

Giffen, C. "Cyclic branched coverings of doubled curves in 3-manifolds." *Illinois J. Math.*, vol. 11, pp. 644–646.

Hosakawa, F. "A concept of cobordism between links." *Ann. of Math.*, vol. 86, pp. 362–373.

Levine, J. "A method for generating link polynomials." *Amer. J. Math.*, vol. 89, pp. 69–84.

Magnus, W. and Peluso, A. "On knot groups." *Comm. Pure Appl. Math.*, vol. 20, pp. 749–770.

Massey, W. S. ALGEBRAIC TOPOLOGY: AN INTRODUCTION. Harcourt, Brace & World, Inc.

Murasugi, K. "Errata to: On the center of the group of a link." *Proc. A.M.S.*, vol. 18, p. 1142.

Schaufele, C. B. "Kernels of free abelian representations of a link group." *Proc. A.M.S.*, vol. 18, pp. 535–539.

Schaufele, C. B. "The commutator group of a doubled knot." *Duke Math. J.*, vol. 34, pp. 677–682.

Smythe, N. "Isotopy invariants of links and the Alexander matrix." *Amer. J. Math.*, vol. 89, pp. 693–703.

Smythe, N. "Trivial knots with arbitrary projection." *J. Australian Math. Soc.*, vol. 7, pp. 481–489.

Index

Cross references are indicated by italics. The numbers refer to pages.